© **Colin Hardwick**

Portions of information contained in this publication/book are printed with permission of Minitab Inc. All such material remains the exclusive property and copyright of Minitab Inc.

All rights reserved.

Minitab Statistical Software

Minitab Statistical Software supports virtually every major Lean Six Sigma initiative around the world. Minitab's unparalleled ease of use, comprehensive methods, and compelling graphics help quality professionals effectively analyze their data and target the best opportunities for business improvement.

This book contains references to Minitab Statistical Software. As a courtesy, readers can download a free 30-day trial at www.minitab.com.

Minitab

Minitab®, and the Minitab logo are all registered trademarks of Minitab, Inc., in the United States and other countries. Copyright ©2011 Minitab Inc. All rights Reserved

Contents

Introduction 4
What is Design of Experiments? 4
When Should It Be Used? 5
Main steps in solving a problem using DoE 5
    Step 1: Identify the Factors, Responses and Levels 5
    Step 2: Create a Design 6
    Step 3: Carry Out the Experiment 16
    Step 4: Analyse the Results 16
A Worked Example 37
    Identification of factors, levels and responses 37
    Creating the Design 38
    Running the Experiment 38
Summary 46
Glossary of Terms 47
Appendix 1: Normal Plots of the Effects for Multilevel Factorial Designs 48

Introduction

This guide has been produced specifically for engineers to use as a practical tool in devising suitable experiments with which to solve real world problems. It does not discuss the history or mathematics of the technique except where absolutely necessary to illustrate an important point.

Design of Experiments is an incredibly powerful tool and, in the right hands, will cut down significantly the normal type of trial and error style experiments found routinely whenever engineers are trying to solve a problem!

In this guide a software package called Minitab® Statistical Software is used to design experiments and analyse the results. You can download a free 30 day trial at www.minitab.com

Minitab is just one software package, of many, that will do this but it is my preferred package simply because I believe it to be the best available – both for ease of use and accuracy of analysis. I have included a number of screenshots taken from Minitab where such screenshots aid explanation.

Traditional teaching of DoE is generally poor in my opinion. This is for two principal reasons:

1. The teaching is carried out by a mathematician or statistician, who blinds the student with complex formulae and makes the subject seem far more complex than necessary. Result - students who will forever decry the technique as too complex or too time consuming to be of any practical use in the real engineering world.
2. The teaching is carried out by a "real world" engineer, who does not have sufficient grasp of the subject matter to convey it succinctly and simply. Result; students who don't believe that the technique actually works and therefore steer clear of it in preference of what they know best – changing one factor at a time and testing the result. I will show later that in the vast majority of cases this is quite simply the wrong approach.

So, work through the text. Don't worry its short; and use Minitab to try out the various examples. You will quickly see just how powerful this technique is in practice and how easy it is to use.

What is Design of Experiments?

So what is "Design of Experiments" or DoE as it is known in its frequently abbreviated form?

Basically it is a controlled experiment in which the experimenter is interested in both quantifying and optimising the effect of a number of input variables on one or more output variables. By quantification we mean estimating the size of the effect of each input variable. Optimising is the process of finding the optimal setting of each input variable which gives a required output.

Let's define two aspects here;

- The input variables causing the variation are called "factors".
- The output variables being measured are called the "responses".

When Should It Be Used?

In engineering and science, the most common reasons for using a DoE are:

- To identify the factors which most affect the responses
- To identify those factors which act **together** on a particular response – these are called "interactions"
- To determine the optimum settings of factors to satisfy a number of required responses

There are a number of other reasons, but let's keep it simple for now and consider only the above reasons. These will account for most of the problems you decide to solve.

Main steps in solving a problem using DoE

Step 1: Identify the Factors, Responses and Levels

The first step is to identify the factors we are going to test.

The initial DoE is simply to quantify the effect of each main factor on the responses and also if any combination of factors act together, in combination, on the response(s).

Gather the experts on the process, and ask them the simple question:

"What factors are most likely to have an effect on the response?"

At this point you are likely to get a relatively large number of factors, and probably some debate and disagreement on the size of effect they might have!

Push the debate hard and drive for a lower number of factors; the ones that are **most likely** to have an effect. The reason for this is very simple; the more factors you test in one experiment, the more runs are required in that experiment. I tend to aim for 3 or 4 factors in the first instance. Each trial in the experiment is called a "run".

Do not be afraid to challenge each factor that is proposed and don't be afraid to offer some of your own. Subject matter experts are often blinded to potential factors which may have a significant effect on the responses simply because they are too close to the subject and fail to take a systems type view of the problem.

Let's assume we decide upon 3 agreed factors.

We now need to agree the "levels" of each factor. This simply means that we will test each factor at a number of settings. Again, the more levels you decide to test, the more runs will be required.

As an example, suppose we have 3 factors at two levels. This means that to test every combination of factors we will need $2^3 = 8$ runs (2 levels$^{\text{3 factors}}$)

Testing every combination means that all of the main effects and interactions can be estimated. The name for this type of design is a *"full factorial"*.

If you have a large number of factors or levels then the number of runs can get too time consuming or expensive. It is possible in these cases to run what are termed *"fractional factorial"* experiments. This simply means that not all combinations are tested. This also means that some of the main effects and interactions cannot be estimated independently – this is known as *confounding.*

In the majority of cases it is better to reduce the number of factors and levels to those which are considered to have the most significance and then carry out a full factorial experiment.

Let's keep it simple and assume we have chosen three factors: A, B and C with two levels which we will call High and Low. We have one response, which we will call Response X.

Step 2: Create a Design

We have the factors, levels and responses identified. It is therefore time to choose a suitable design.

To do this:

Enter Minitab and click on **Stat/DoE/Factorial/Create Factorial Design**

You will then see this:-

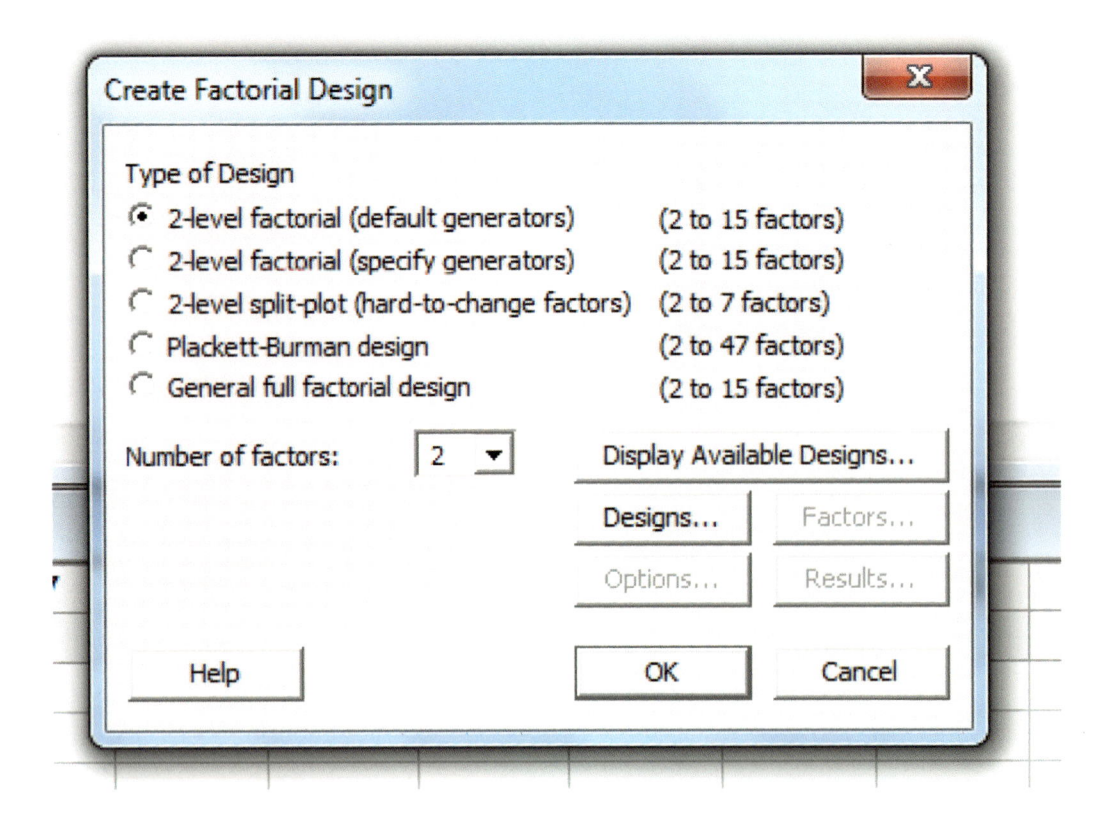

You'll see that you have a choice of designs. Let's look at the options and what they mean.

The only difference between the first two options is that one uses "default generators" and the other requires you to "specify generators". But what does this mean and is it important? The short answer is that for full factorial designs (which I recommend) **they are not important**. This is because the generators simply determine what subset of experiments is selected from the full factorial set of runs to form a fractional factorial set of runs. Since we are using a full factorial design then the generators are not relevant.

The third option is a 2 level split-plot (hard to change factors). This type of design is used when there is at least one factor that is difficult to randomize. This is normally due to the fact that randomizing such a factor is expensive or time consuming. In reality it is rarely necessary to use such designs but if you do wish to use such a design then consult a six sigma Master Black Belt or Black Belt.

The fourth option is Plackett-Burman designs. These were developed in an attempt to reduce the number of runs required in experiments (such as where the runs are very expensive or time consuming) – the basic problem is that in such designs the interactions between factors is considered negligible. In practice, this is very rare for most engineering experiments. A full factorial experiment, with strictly limited numbers of factors and levels, means that Plackett-Burman designs are rarely required.

The final option is a "General full factorial design". These are used when at least one factor has more than two levels. This can be quite common but clearly using more than two levels will increase the number of runs required. My recommendation is to try not to use **more** than two levels for **more** than one factor since the number of runs will start to increase quite rapidly.

As an example, consider a three factor experiment, where two of the three factors have 4 levels and the remaining factor has two levels. This will result in an experiment with 32 runs:

$$(4 \text{ levels}^{2 \text{ factors}}) \times (2 \text{ levels}^{1 \text{ factor}}) = 32 \text{ runs}$$

You'll remember from earlier that we have chosen three factors: A, B and C with two levels, which we will call High and Low. We have one response, which we will call Response X.

This will result in 8 runs. Again we will keep it simple and carry out each trial only once (we'll discuss replicating each run later).

Click on the button marked "Display Available Designs". You will see this:-

Create Factorial Design - Display Available Designs

Available Factorial Designs (with Resolution)

Run	2	3	4	5	6	7	8	9	10	11	12	13	14	15
4	Full	III												
8		Full	IV	III	III	III								
16			Full	V	IV	IV	IV	III	III	III	III	III	III	III
32				Full	VI	IV	IV	IV	IV	IV	IV	IV	IV	IV
64					Full	VII	V	IV	IV	IV	IV	IV	IV	IV
128						Full	VIII	VI	V	V	IV	IV	IV	IV

Available Resolution III Plackett-Burman Designs

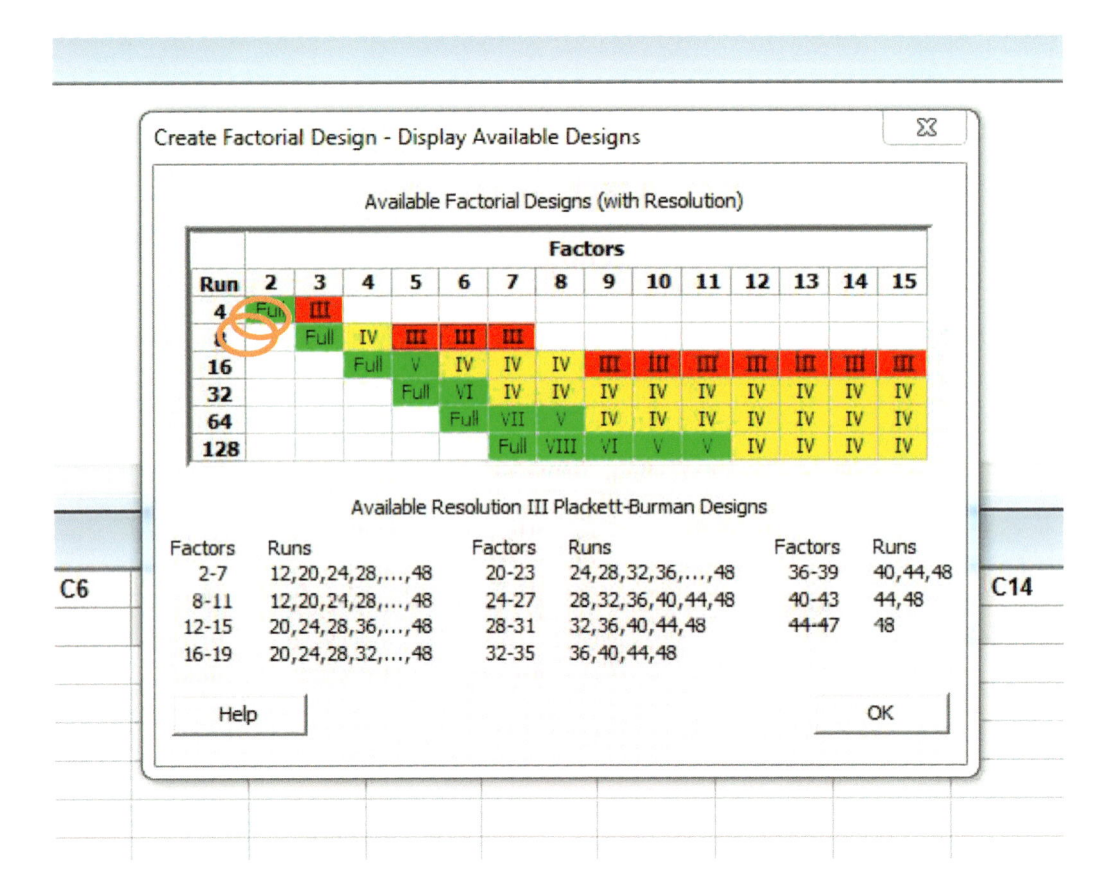

Factors	Runs	Factors	Runs	Factors	Runs
2-7	12,20,24,28,...,48	20-23	24,28,32,36,...,48	36-39	40,44,48
8-11	12,20,24,28,...,48	24-27	28,32,36,40,44,48	40-43	44,48
12-15	20,24,28,36,...,48	28-31	32,36,40,44,48	44-47	48
16-19	20,24,28,32,...,48	32-35	36,40,44,48		

Help   OK

You can see that this is a table of the number of runs in an experiment plotted against the number of factors. In this case we have 3 factors so we are looking at the 2nd column. There are clearly two experiments available; one with 4 runs, which is coloured red and marked resolution III, and the other coloured green and marked "Full".

So what does this mean? Earlier we said that if some of the main effects and interactions cannot be estimated independently then this is known as confounding. This is only a problem in fractional factorial designs because in such designs we are not trying all the possible combinations of factors.

As an example, let's suppose we carry out a fractional factorial design and the main factor A is confounded with the two way interaction BC. This means that B and C act together – i.e. they interact. Such a two way interaction is written as B x C or BC for short.

The overall effect can then be estimated by adding together the effect of A and the effect of BC. Such effects are said to be *aliased*. Effects that are aliased, or confounded, cannot be estimated separately from one another.

Ok, so back to the selection of a suitable design and the confusing term "resolution". This simply means to what extent effects are aliased with other effects in a particular design. **The higher the resolution the fewer the aliased effects and the more you can estimate the effects independently of each other**.

You can see that this only becomes a problem in a fractional factorial design since full factorial designs have no confounding or aliasing – all the effects can be estimated independently of each other.

If you do choose a fractional factorial design, then it is better to choose the one with the better resolution. For example choose a resolution IV in preference to a resolution III.

A resolution IV means that the main effects are confounded with 3 way interactions and a resolution III means that the main effects are confounded with 2 way interactions.

My advice, if you are designing a fractional factorial design, is to always choose an experiment which has high resolution – and therefore coloured green on the table above. These are at least resolution V.

In our case we are running all possible combinations in eight runs, so no confounding or aliasing exists.

The next step is to select the full factorial experiment. To do this, click on the button marked "Designs". You will see this:-

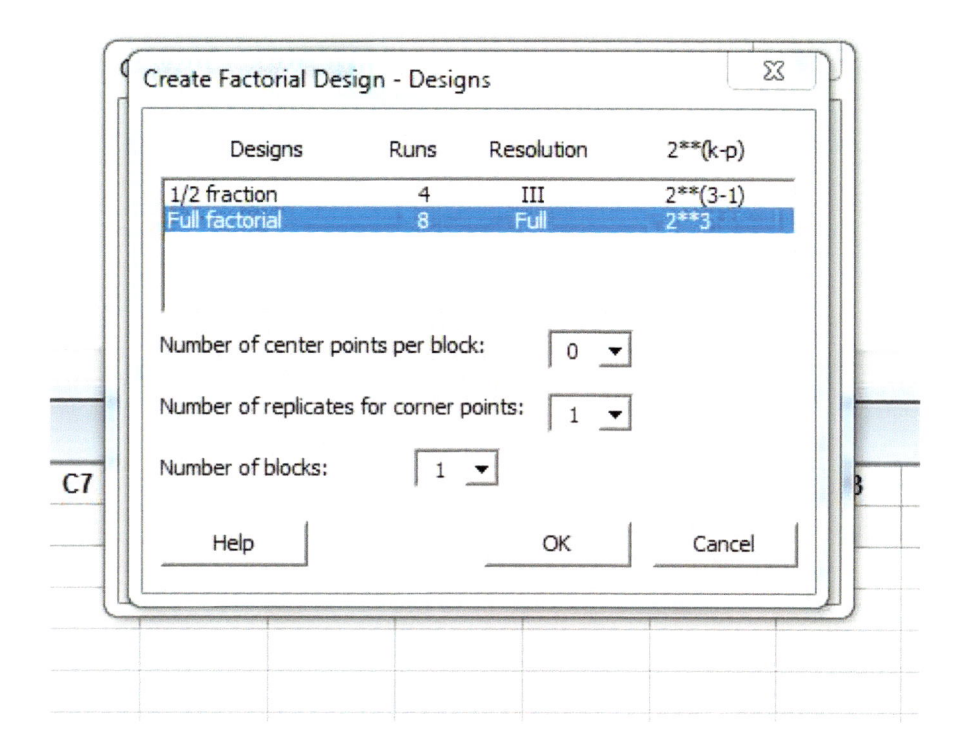

Select the full factorial experiment so that it is highlighted blue and then click "OK".

The other options are:

- Number of center points per block
- Number of replicates for corner points
- Number of blocks.

Let's look at each of these in turn so that we understand them.

Number of center points per block:-

Center points – these are the points at which all the factor levels are set exactly half way between your chosen high and low settings. So, if you had chosen to set the low level for a particular factor at, say for example, 50 units and the upper level for the same factor at, say for example, 100 units then the center point would be at 75 units.

Clearly center points can only be set for numeric factors. Minitab can cope with the fact that you may have some text factors (i.e. non numeric), simply by creating pseudo center points. You can read more about this in the help facility in Minitab but I wouldn't worry too much about it. The only real point is; do you want to have center points in the experiment?

My experience is that for the vast majority of experiments you tend to be trying to quantify the size of effect of each factor and therefore adding center points will only serve to increase the number of runs in the experiment, without providing further insight. I find it best to leave this set to zero until you understand the size of the effect of each factor and have reached the point where you want to optimise these factors.

Number of replicates for corner points:

Corner points – these are simply the experimental runs where all factors are set at either lowest or highest level.

As an example, consider the simplest experiment with two factors at two levels. Let's call the factors A and B and the levels are high and low.

The experimental runs will be:

Run 1; Factor A set to High, factor B set to High

Run 2; Factor A set to High, Factor B set to Low

Run 3; Factor A set to Low, Factor B set to Low

Run 4; Factor A set to Low, Factor B set to High

The corner points are therefore shown as follows:-

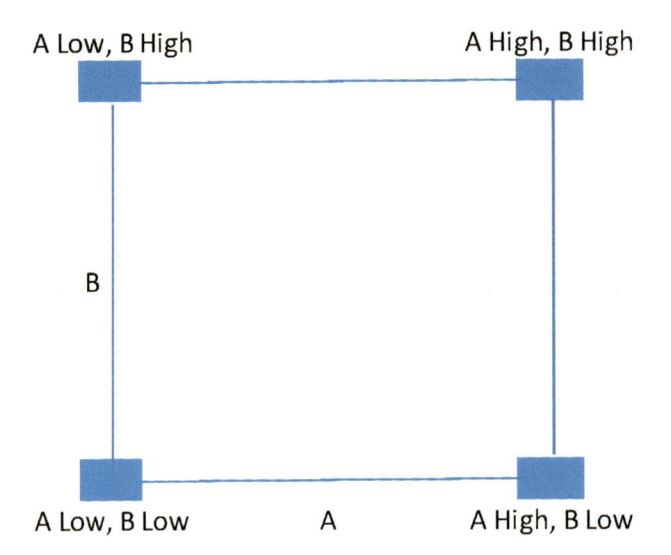

```
    A Low, B High              A High, B High
         ■───────────────────────────■
         │                           │
         │                           │
      B  │                           │
         │                           │
         │                           │
         ■───────────────────────────■
    A Low, B Low        A       A High, B Low
```

The center point will clearly lie at the centre of this box, with both A and B set at mid-point.

So what does "Number of replicates for corner points" mean and is it important?

The term replicates simply means multiple experimental runs with the same factor settings. You may wish to replicate the experiment to estimate the variance at a given setup. In the example above you may choose to carry out the 4 runs once at each of the factor settings or you may choose 2 or more times at each of the factor settings. One experiment with 4 runs is one replicate, repeating the runs again gives 8 runs or two replicates, repeating the runs for a third time gives 12 runs or 3 replicates.

When thinking about using replicates, consider using them where:

- You may be trying to develop a prediction model, in which case using replicates may increase the precision of the model.
- The cost of experimentation using replication is acceptable.

I recommend using replicates if the cost is not prohibitive, but don't worry too much if not.

Number of blocks

In an ideal world an experiment would be carried out under consistent conditions with no other variables other than those being varied in the experiment. This is, of course, virtually impossible in the real world. Blocks are therefore used to minimize experimental error from factors which are not deliberately part of the experiment and may be out of your control.

As an example, for an experiment of any length, the runs may typically take more than one day. This means that other sources of variation may be introduced into the experiment simply because it is spread over multiple days. One example of such sources of variation is the weather (pressure, temperature, wind speed and direction etc). Another example is the change of equipment between different runs.

You may not know what the other sources of variation are, but simply that they may exist.

These potential differences can be accounted for by using a blocking variable. This means you can clearly distinguish between variation caused by the experimental factors and those due to the blocking effect. For example, a blocking variable we may use is "day", to account for the variation in results due to running the trial over more than one day.

If you click on the arrow next to "number of blocks", you will see a list of options. For example, if you are carrying out a 2 factor, 3 level experiment with one replicate then you will have a choice of either:

- 1 block (all runs analysed as one block and taking no account of the day variable),
- 2 blocks (runs taking place as two blocks and therefore taking account of the day variable, or
- 4 blocks (runs taking place as four blocks and therefore taking even more account of the day variable, maybe splitting it into morning and afternoon shifts)

Note that changing the number of replicates will change the number of blocks available. In this case, using two replicates will give a further option of 8 blocks. Note also that this will increase the number of runs within the experiment.

The number of blocks is an option – my advice is to use it only when necessary and when you suspect that the sources of variation outside the experiment are significant in size when compared with the sources of variation within the experiment. If not then leave the number of blocks set to 1.

Let's continue with creating the design. Assume we have chosen to have no center points, one single replicate and one block because we can carry out the experiment on one day and suspect that the main sources of variation are contained within the experiment.

Click OK on the screen shown below:

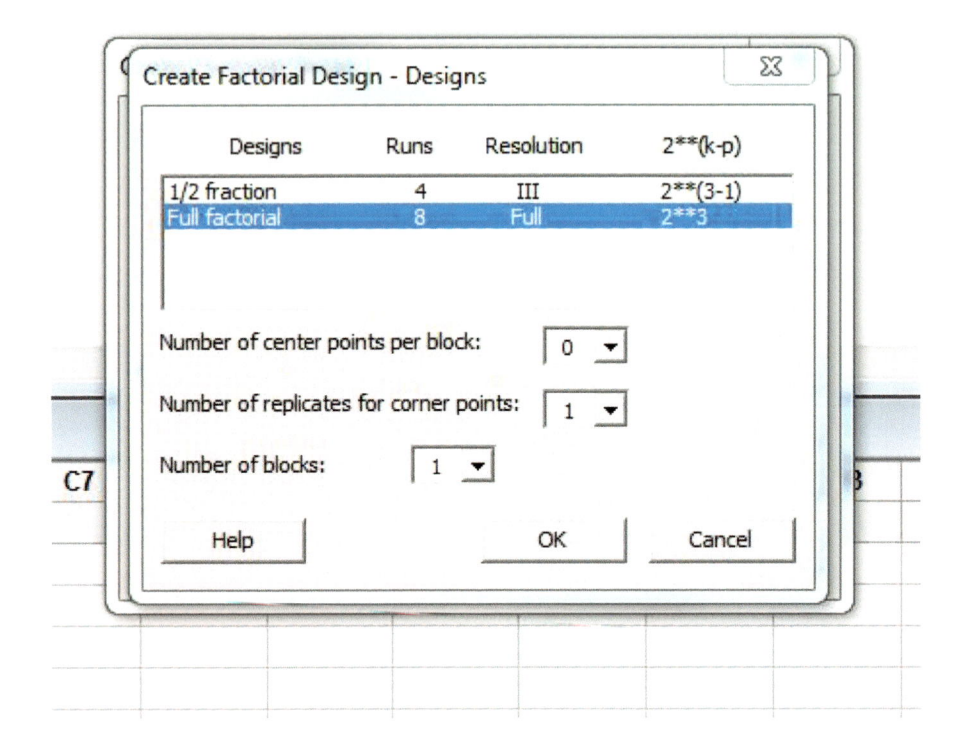

Designs	Runs	Resolution	2**(k-p)
1/2 fraction	4	III	2**(3-1)
Full factorial	8	Full	2**3

Number of center points per block: 0

Number of replicates for corner points: 1

Number of blocks: 1

This will take you back to this screen:

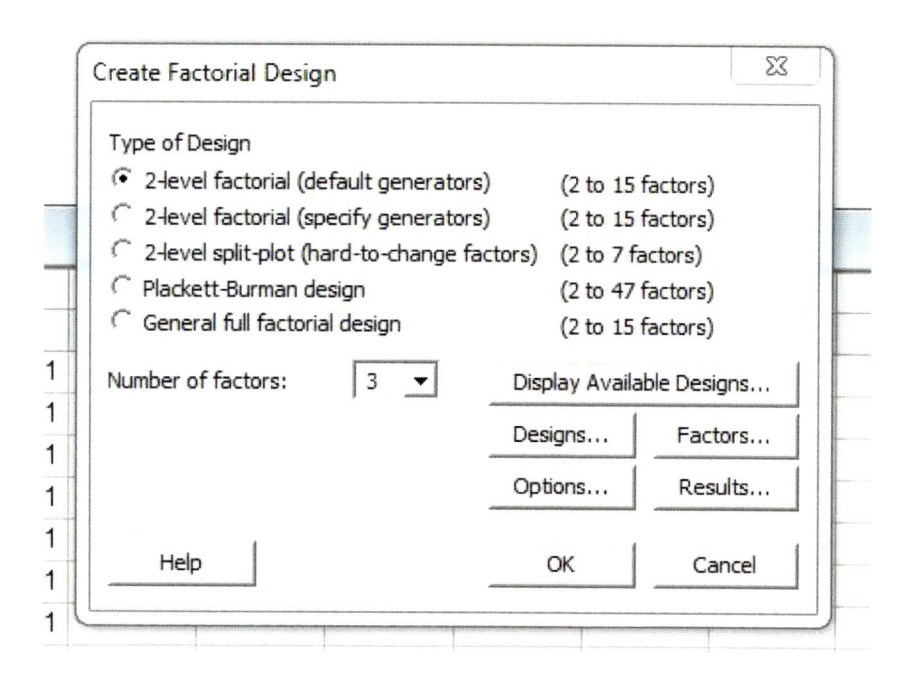

The next step is to click on the button marked "Factors". You will then see this screen:-

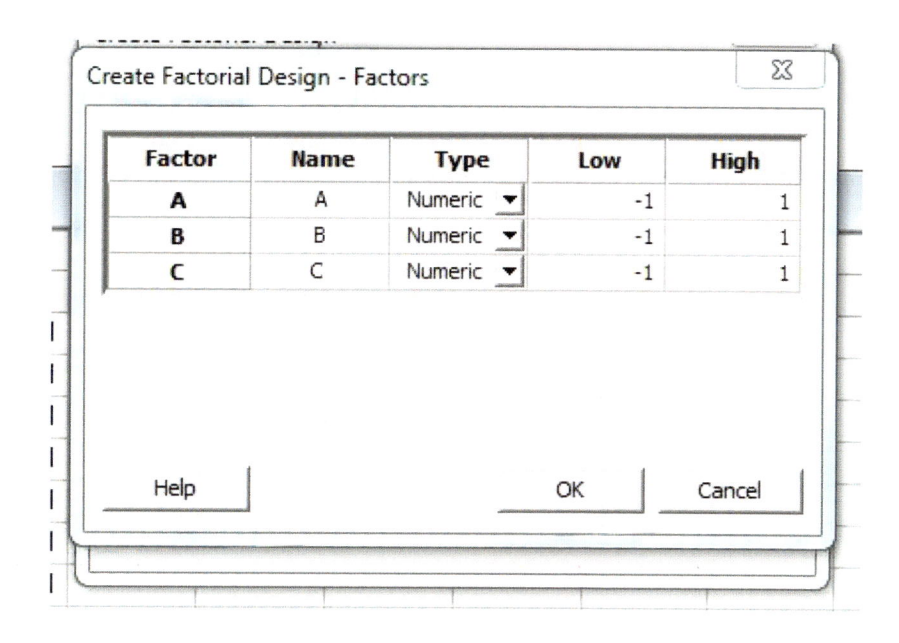

Each factor can now be given; a name, the type specified (Numeric or text), and the values of the low and high levels named; simply by typing into the relevant boxes.

As an example, let's assume that we have a chemical mixing experiment, with the factors being temperature, speed and the actual mixing vessel being used. The response of interest is the viscosity of the resulting mix. **In this case, we are looking for the factors which will give you a viscosity as low as possible at minimum cost.**

Enter the factors into the experiment as follows:-

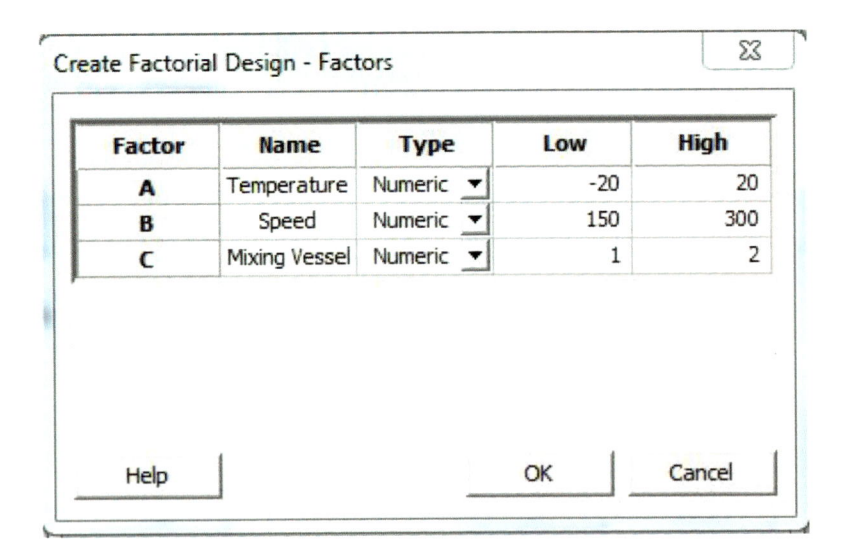

Create Factorial Design - Factors

Factor	Name	Type	Low	High
A	Temperature	Numeric	-20	20
B	Speed	Numeric	150	300
C	Mixing Vessel	Numeric	1	2

Click OK and you will be back at the previous screen. Now click on "Options". You will see this screen:-

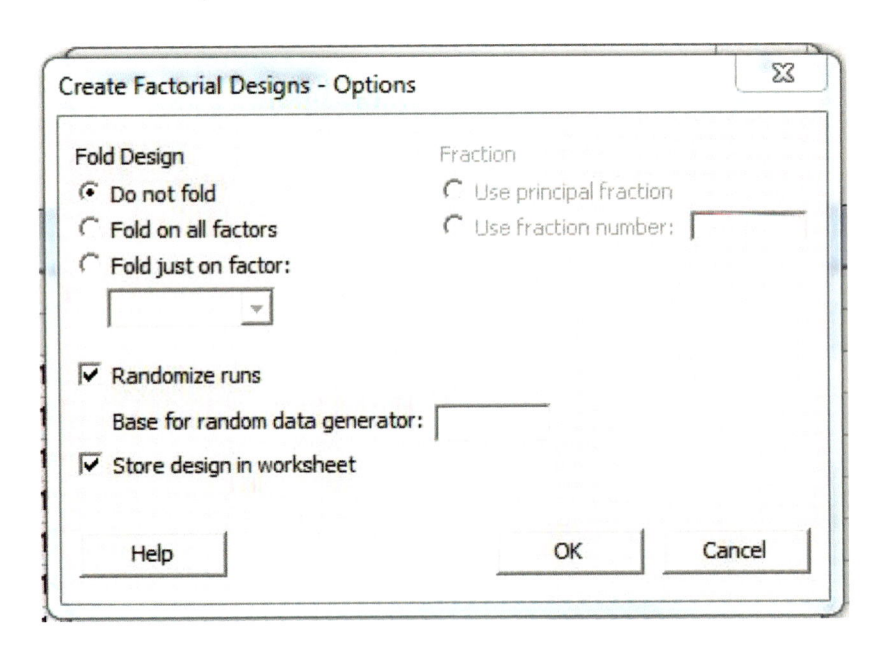

Create Factorial Designs - Options

Fold Design
- Do not fold
- Fold on all factors
- Fold just on factor:

Fraction
- Use principal fraction
- Use fraction number:

☑ Randomize runs
Base for random data generator:
☑ Store design in worksheet

You will see a choice of fold designs and both the "Randomize runs" box and the "Store design in worksheet" will be ticked automatically.

Folding is a way to reduce confounding (you remember what that is, right?). For full factorial experiments, there is no confounding and therefore this is not relevant. If you have a full factorial experiment you can therefore leave this set to "Do not fold".

Randomizing is used to lessen the effects of factors outside the experiment (i.e. not included) and therefore it is usually a good idea to do so. Hence the box is ticked automatically.

Minitab shows both Standard Order (non-randomized) and Run Order (randomized) for the experiment; in columns C1 and C2 respectively. If you do not randomize then both Standard Order and Run Order will be identical.

"Store design in worksheet" means exactly that – Minitab stores the design. This is always a good idea if you want to refer back to the design itself and also if you are trying a number of designs to see which suits your needs the best.

So, always randomize the design and carry out the experiment in the run order if possible. If not then don't worry too much about it but understand that you should try and ensure that the effect of the factors within the design need to be much larger than the effect of the factors outside the experiment if you don't randomize.

Click "Ok" and the experiment will be created. You'll see something like this:-

↓	C1	C2	C3	C4	C5	C6	C7
	StdOrder	RunOrder	CenterPt	Blocks	Temperature	Speed	Mixing Vessel
1	8	1	1	1	20	300	2
2	6	2	1	1	20	150	2
3	2	3	1	1	20	150	1
4	4	4	1	1	20	300	1
5	1	5	1	1	-20	150	1
6	7	6	1	1	-20	300	2
7	5	7	1	1	-20	150	2
8	3	8	1	1	-20	300	1

You can see that C1 is reserved for the Standard Order, C2 for the Run Order, C3 for the Center points, C4 for the blocks, and then as many columns as required for the factors. In this case, there are three factors; temperature, speed and mixing vessel.

Ok, that's it!

You have designed an experiment. In reality, this has taken far longer to explain here than it takes in practice. It's easily possible to design an experiment in a few minutes once you get used to doing it.

Let's move onto step 3 which is actually carrying out the experiment.

Step 3: Carry Out the Experiment

The trick here is to make sure that the experiment is carried out exactly as designed and ideally in the run order provided.

It's very common to be asked to allow the experiment to be carried out in a practical order rather than in a randomized order.

As an example in the above case, you might be asked if all the runs at -20°C can be carried out first, followed by all the runs at +20°C.

You can allow this, but be aware that what you are doing is taking away the randomization of the run order.

Remember from step 2 that;

"Randomizing is used to lessen the effects of factors outside the experiment (i.e. not included)"

So before you allow the non-randomization, satisfy yourself that taking away the randomization will not adversely affect the results. For example, especially in cases where you suspect that factors outside the experiment might have a significant effect.

If it's possible, carry out the experiments on one day, or one shift, to minimize outside effects.

The other significant issue to consider is the accuracy and precision of measurement when carrying out the experiment. All input factors and all responses MUST be measured to an appropriate degree of accuracy and precision. If they are not then the measurement error will mask the actual variation due to different settings of the factors themselves.

So how do you know when you have sufficient accuracy and precision of measurement? The simple answer is that the accuracy and precision of measurement should be much smaller than the variation you expect to see in the experiment. If you have an accuracy of, say, 10 units, and the variation expected from the experiment is also 10 units; then clearly you will not be able to tell what is causing the differences in measured value – is it the accuracy/precision of measurement or, one or more of, the factors?

Step 4: Analyse the Results

Let's assume that the results are now in and we have them in front of us.

How do we enter then into the experiment in Minitab?

Well, it's actually very easy. You simply place them into the column immediately to the right of the factors!

We said earlier that we are only interested in one response, which is the viscosity of the resulting mixing operation. We enter the results into column C8, labelled "Viscosity".

In this case the results would look something like this:-

↓	C1	C2	C3	C4	C5	C6	C7	C8
	StdOrder	RunOrder	CenterPt	Blocks	Temperature	Speed	Mixing Vessel	Viscosity
1	8	1	1	1	20	300	2	18
2	6	2	1	1	20	150	2	19
3	2	3	1	1	20	150	1	16
4	4	4	1	1	20	300	1	17
5	1	5	1	1	-20	150	1	15
6	7	6	1	1	-20	300	2	18
7	5	7	1	1	-20	150	2	17
8	3	8	1	1	-20	300	1	15

We can now go ahead and start to analyse the results.

Firstly, we define the responses that we are interested in; and in this case we have only one response which is "viscosity".

To define this, click on **Stat/DOE/Factorial/Analyse Factorial Design**

You'll see this input screen:-

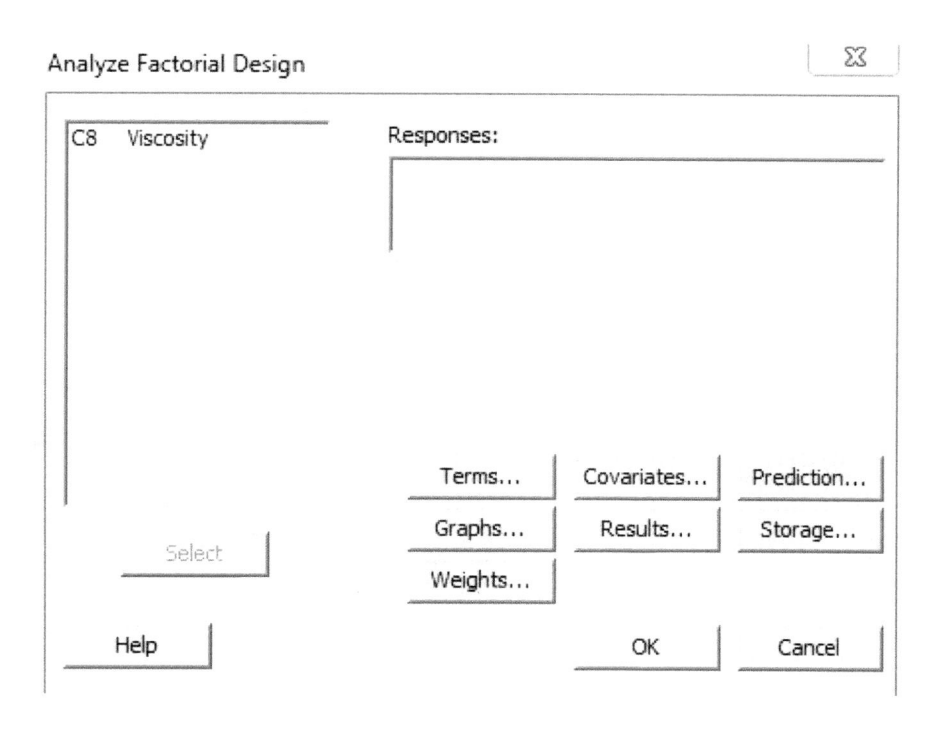

Click on the word "viscosity", so that it is highlighted blue, and then click on the "Select" button. The response (viscosity) will now appear in the responses table and look like this:-

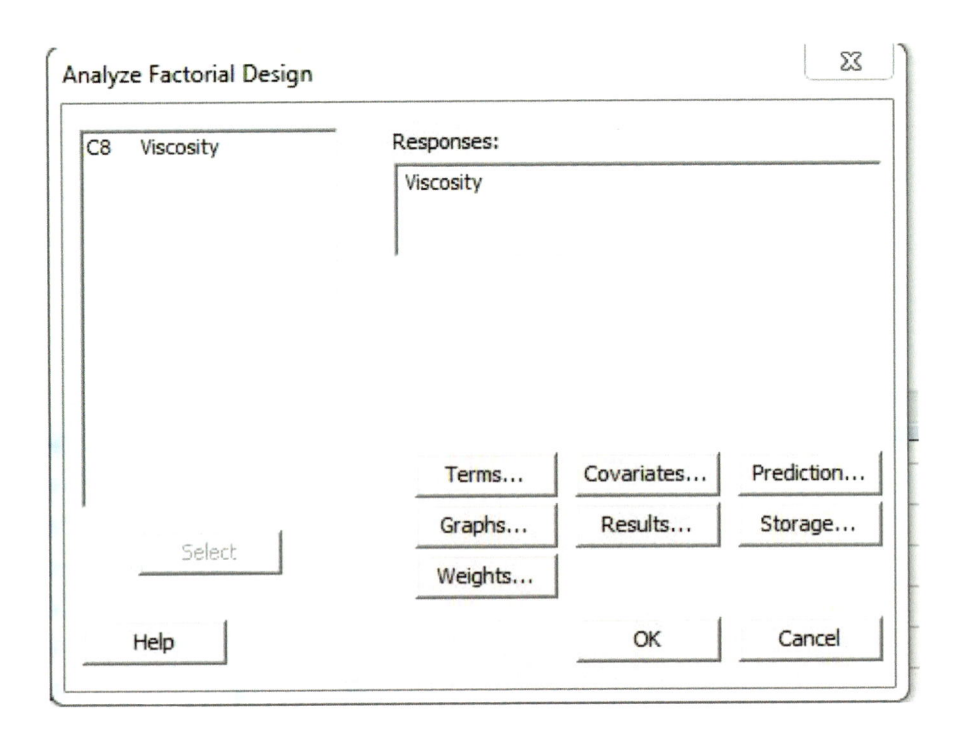

We'll now work through the various options and what they mean.

Terms: this allows you to choose which terms to include in the model. The terms are the main effects and their interactions. They could be 2 way interactions (2 factors acting together), 3 way interactions (3 factors acting together) or higher level interactions.

Click on the button marked "Terms" and you'll see this input screen:-

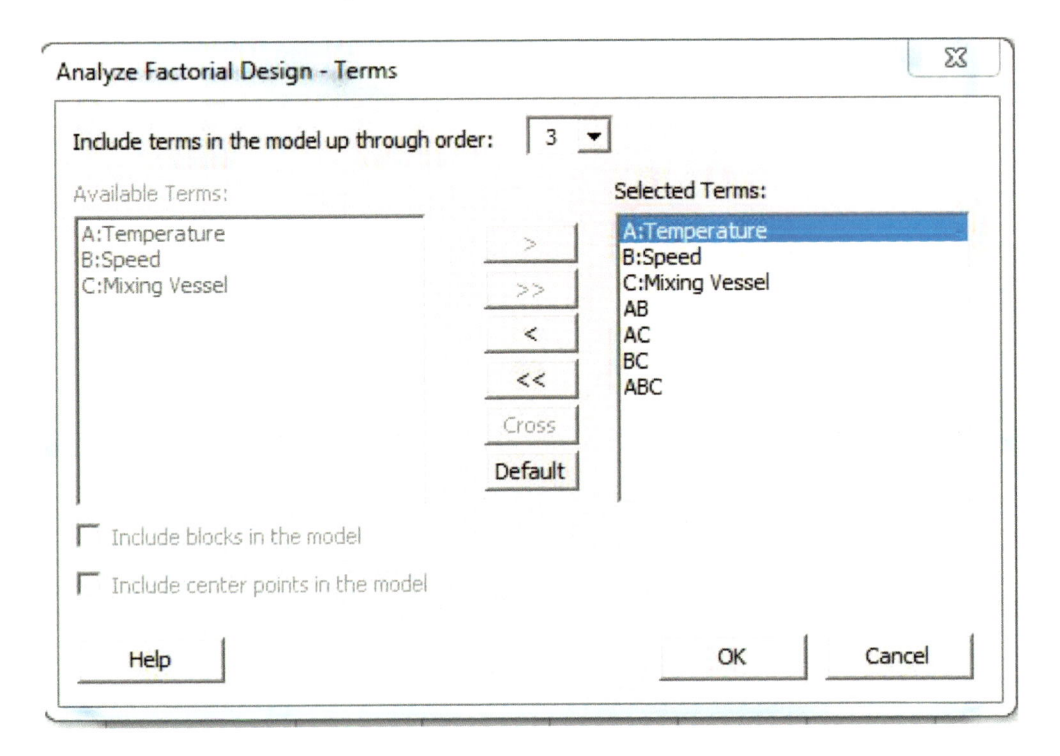

You can see that there is a pull down box just to the right of the words "Include terms in the model up through order:"

In this case we have 3 main effects so the maximum order is 3. This means we can look at all main effects, all two way interaction effects and all 3 way interaction effects. Minitab sets the default to the maximum number since it assumes you want to look at all main effects and all interaction effects.

If you wish to look at just the main effects with the two way interaction effects, then the number should be changed to 2.

If you wish to look at just the main effects with no interaction effects, then the number should be changed to 1.

We will leave this set to 3, since we are interested in all main effects and all interaction effects.

Covariates: - The next button is marked "Covariates". A covariate is used to account for the effect of a particular variable which is difficult to control. In this case we can control all factors easily and therefore we have no covariates. You will find that in the majority of engineering experiments it is possible to omit covariates by making sure that the factors chosen can be controlled easily.

Prediction: - This allows you to calculate predicted response values for new runs containing different levels for particular factors. I have found this of limited value when carrying out experiments as I describe in this guide since I prefer to get an indication of the size of each effect and then optimise those effects with a further experiment. I can normally then produce a predictive model from those further experiments.

Graphs: - The whole point of these types of experiments is to identify the factors that have a significant effect on the responses. Ideally we want there to be a small number of these since that means there is less to control to keep the response at the value we want.

Click on "Graphs" and then you will see this input screen:-

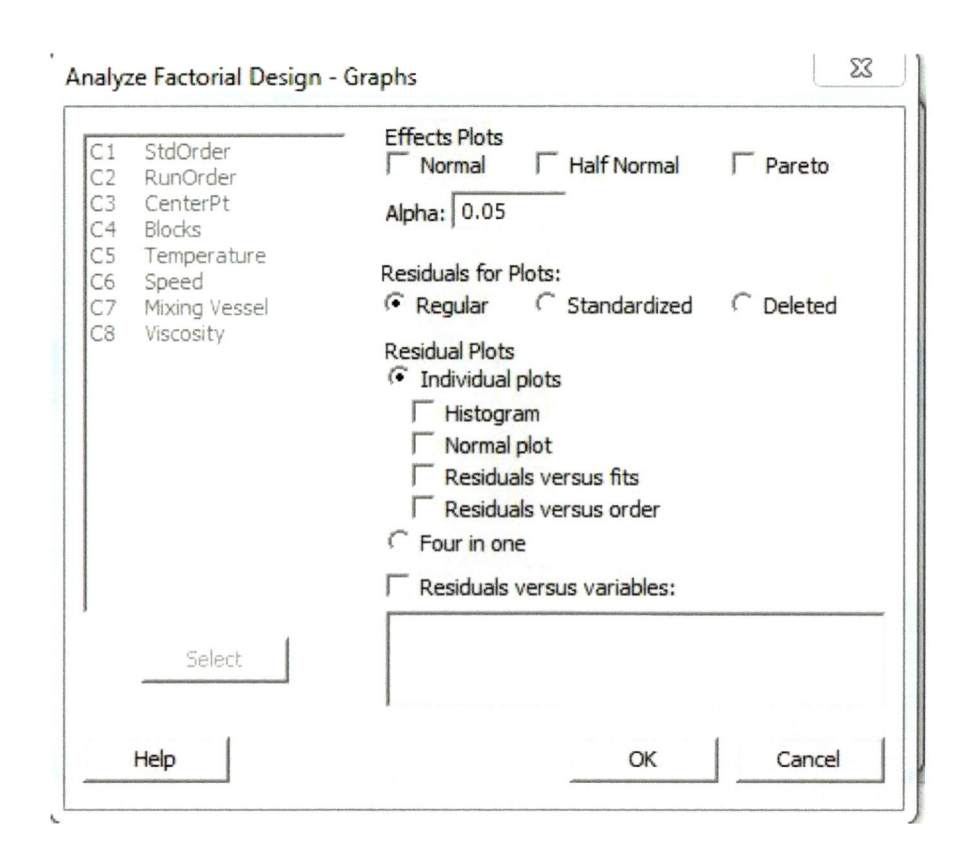

There is a choice of three types of graph: the normal plot, the half normal plot, and the Pareto chart. Using these we can compare the magnitudes of the effects against each other and also see if they are statistically significant.

The "Alpha" value is simply a measure of the amount of confidence you can have in the results. It will be a number between 0 and 1. The lower the number the more confidence you can have in the results and the conclusions you draw from those results. The normal default value is 0.05, which means you can be 95% confident that the results are real and not the result of chance.

Now, under "Effects Plots", select "Normal" and "Pareto". I have found that selecting "Half Normal" to be of very limited value since it tells you virtually the same information as a normal plot.

Under "Residual Plots" select "Four in one". Now select "Residuals versus variables".

You should have a screen which looks like this:-

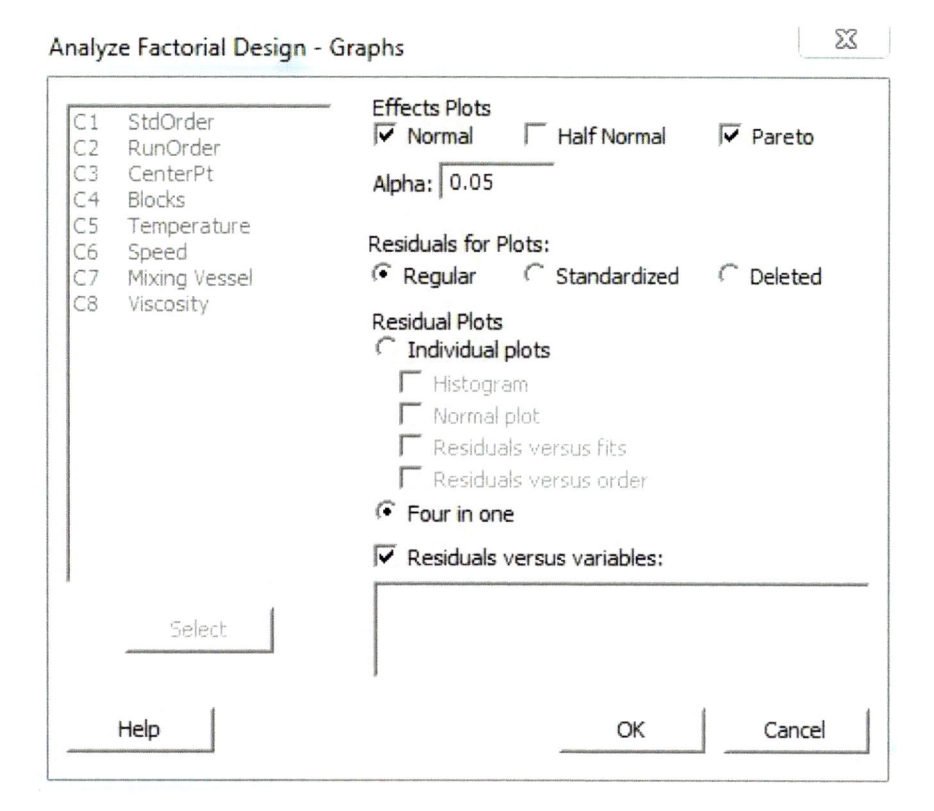

Click "OK" to go back to the options screen.

Results: - This allows you to choose what you want to see in the session window output and it is rarely necessary to change any of the default settings. The screen looks like this:-

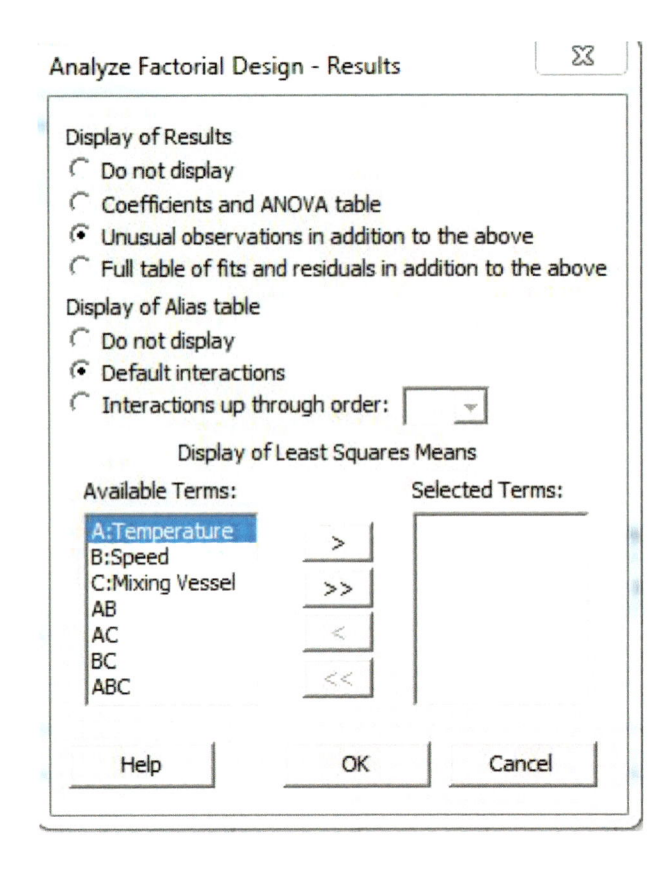

Click "OK" to go back to the options screen. For completeness, here is an explanation of the various options:

Storage: - This allows you to select a number of parameters to assist in further analysis. Selecting any option will result in Minitab storing your selections in the next available column and automatically naming the column. In practice, I have rarely needed to use this option and I recommend you ignore it unless you have some further analyses in mind.

Weights: - This option allows you to weight your model, and perform weighted regression. In thirty years I've never found a need to call on this option!

Click "OK" to go back to the options screen and then click "OK". Minitab will then produce the plots you specified. In this case we selected the normal plot, the half normal plot, and the Pareto plot for the effects.

Let's look at what the two selected plots tell us.

The Normal Plot for effects – allows you to compare the magnitude and statistical significance of the effects, including the interaction effects.

Minitab draws a straight line and the further away from the straight line a point appears the more significant the effect represented by that point.

If any effects are significant then Minitab colours them red and labels them. This will be done for the default alpha value of 0.05 unless you specified a different value earlier.

In this case the plot looks like this:-

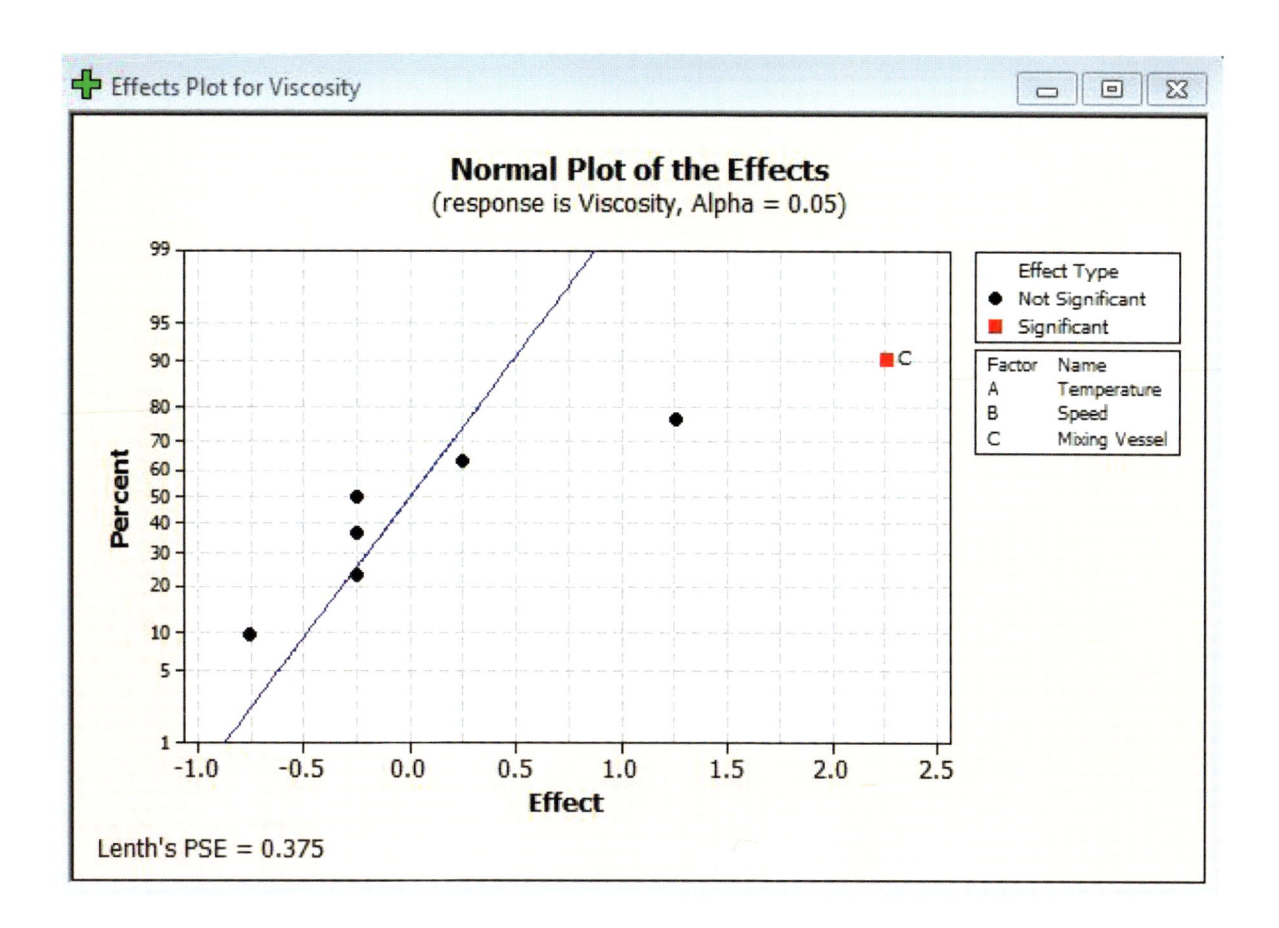

Effects Plot for Viscosity

Normal Plot of the Effects
(response is Viscosity, Alpha = 0.05)

Effect Type
- Not Significant
- Significant

Factor	Name
A	Temperature
B	Speed
C	Mixing Vessel

Lenth's PSE = 0.375

You can see that the only statistically significant effect at the chosen alpha value of 0.05 is effect C - "Mixing Vessel".

The Pareto plot for effects – this again shows you the statistically significant effects, in order of significance. The plot in this case looks like this:-

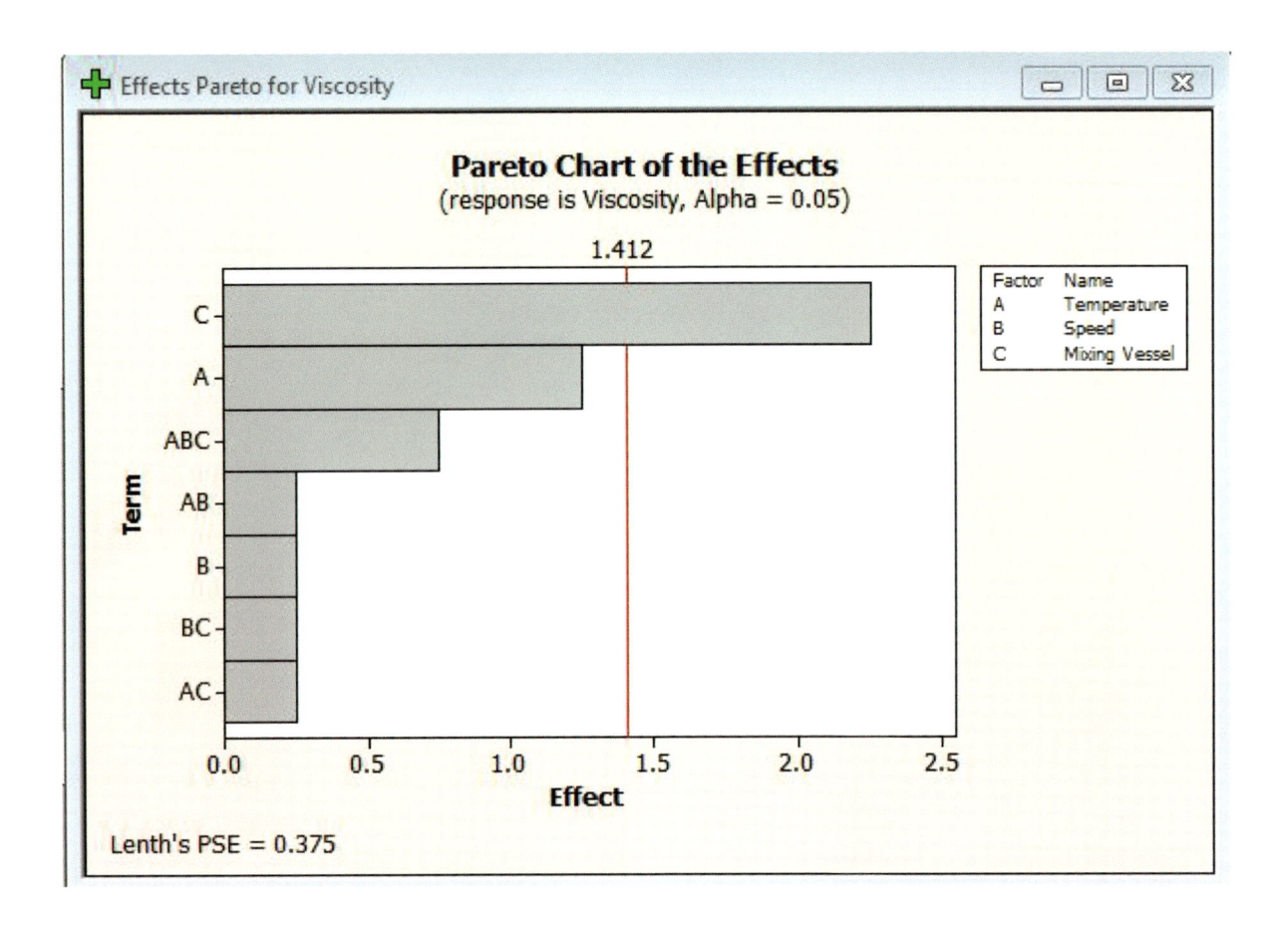

Pareto Chart of the Effects
(response is Viscosity, Alpha = 0.05)

Factor	Name
A	Temperature
B	Speed
C	Mixing Vessel

Lenth's PSE = 0.375

You can see that C is by far the most significant factor, but that A is coming close to being significant at the chosen alpha level.

This is where some judgement is needed.

Just because an effect is not significant at the alpha level of 0.05, this does not mean it should necessarily be ignored. You may decide that since it is easy to control and it could be statistically significant at a slightly lower level of confidence (i.e. a higher alpha level) then you will continue to assess it in the optimising experiments that we will consider later.

One useful action is to go back and drop the alpha level and see if the factor becomes significant. Let's do this for this particular case.

Click on **Stat/DOE/Factorial/Analyse Factorial Design** and then click on "Graphs"

Now change the alpha level to 0.1 rather than 0.05. This means you want to assess the results at 90% confidence rather than 95% confidence. You should have an entry screen that looks like this: -

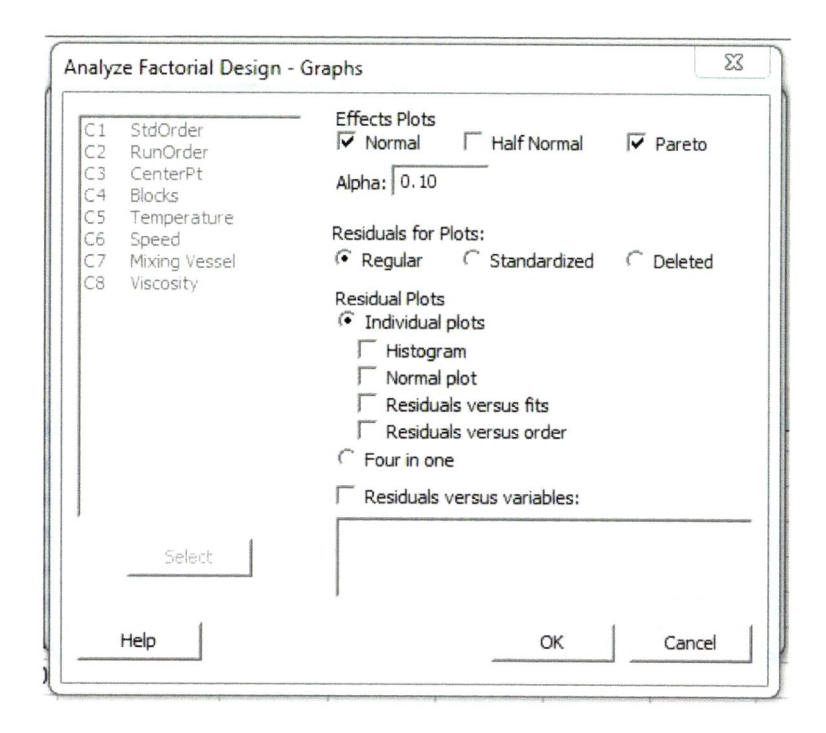

Now Click "OK" and then "OK" again on the following screen. Minitab will plot the graphs using the newly defined alpha level. They should look like this:

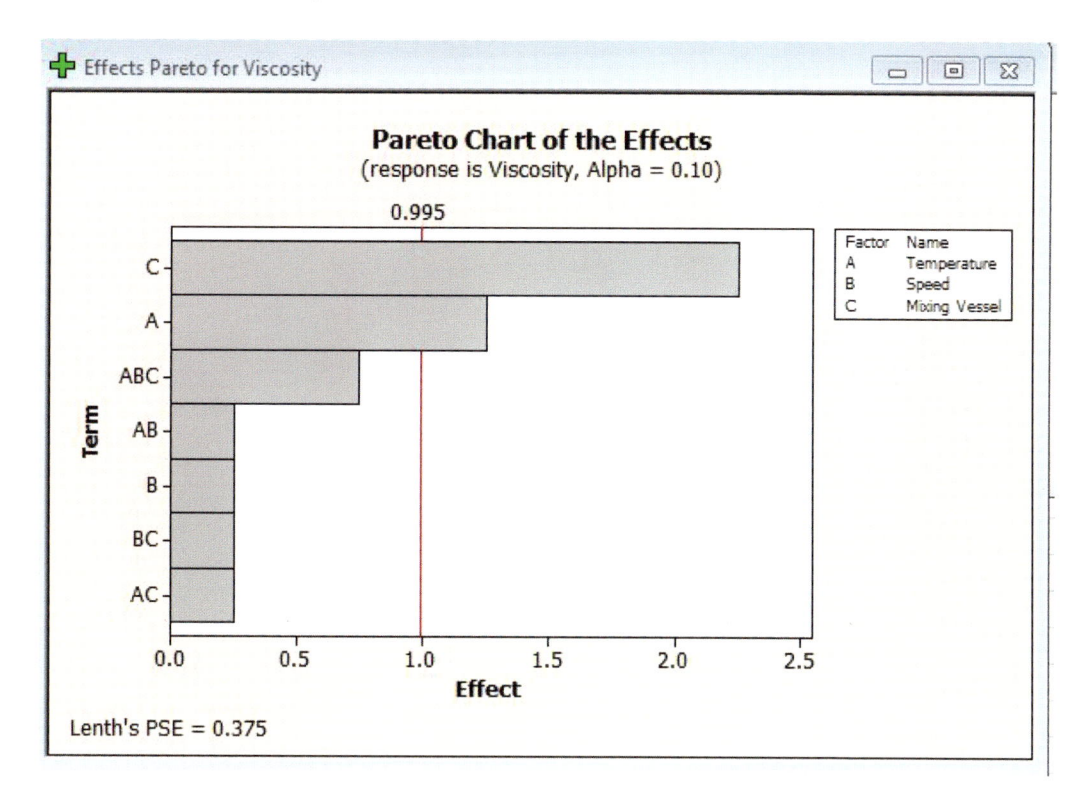

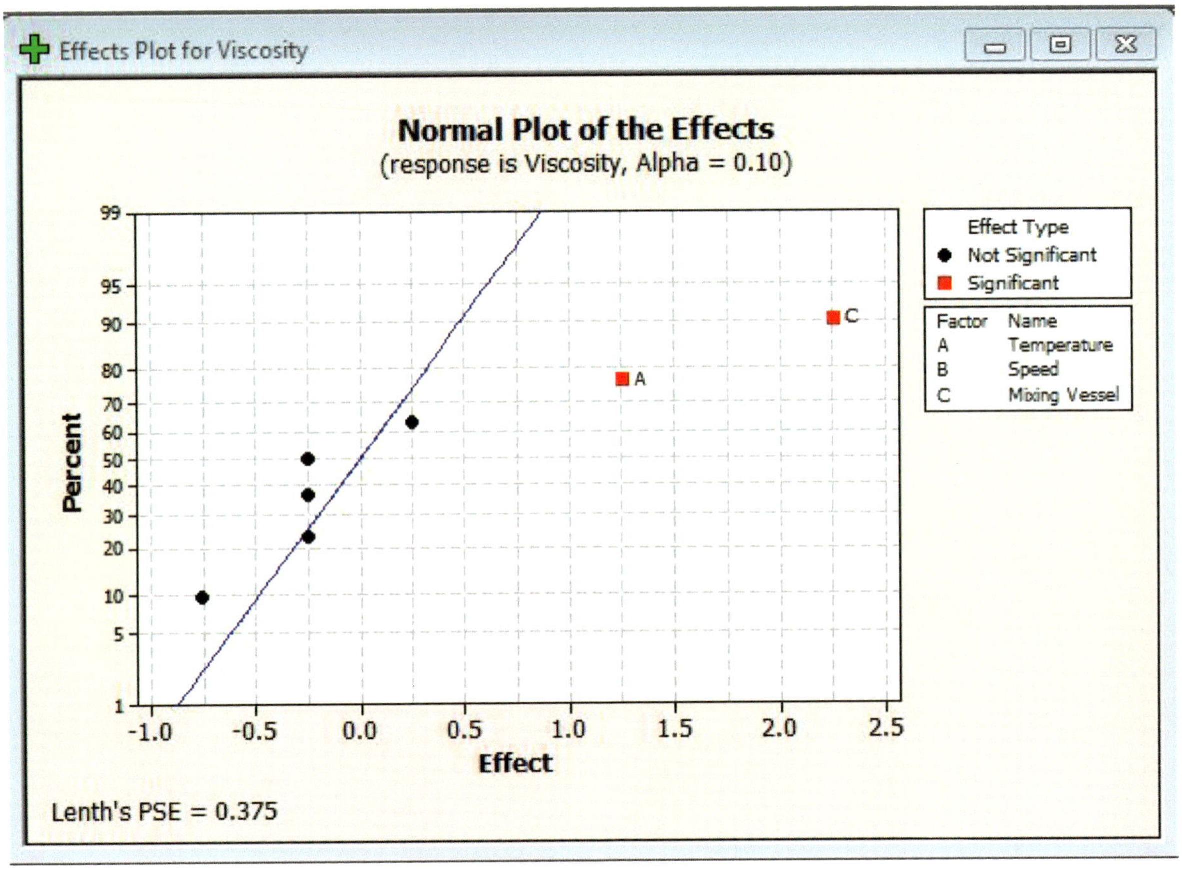
Effects Plot for Viscosity
Normal Plot of the Effects
(response is Viscosity, Alpha = 0.10)
Percent
Effect
Effect Type
Not Significant
Significant
Factor Name
A Temperature
B Speed
C Mixing Vessel
Lenth's PSE = 0.375

You can see that factor A, which is temperature, has now become significant. You may decide that you wish to include temperature in the follow-on experiments since you consider it is significant enough to be worthy of further consideration.

Let's assume that we have decided to consider temperature. The next step is to take a look at the main significant effects, which in this case are A and C (temperature and mixing vessel).

Click on **Stat/DOE/Factorial/Factorial Plots**

You'll now see this input screen:-

[Factorial Plots dialog box shown]

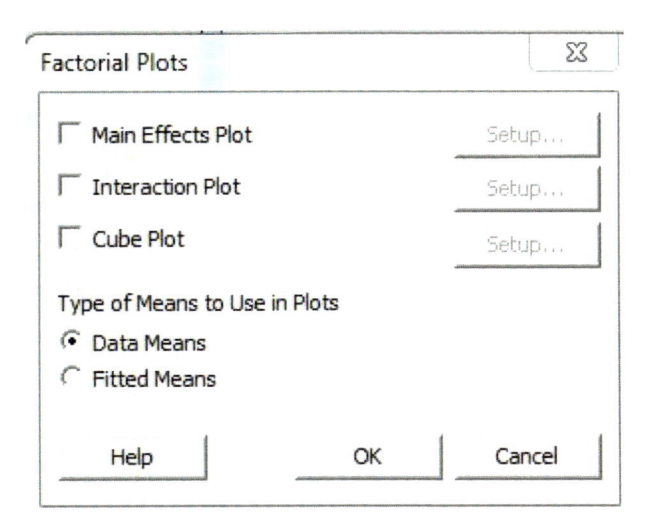

Tick the boxes marked "Main Effects Plot" and then click on the box marked "Setup". You'll see this:

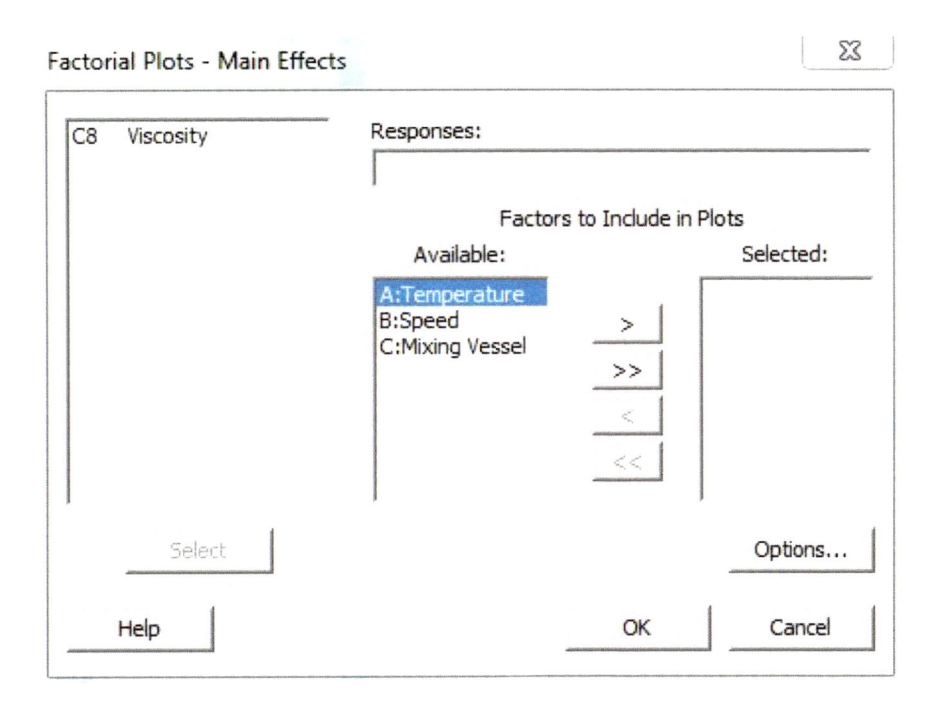

[Factorial Plots - Main Effects dialog box shown]

We can now enter the responses and the factors to include in the plots. Select C8 "Viscosity" and click "Select". Choose both temperature and mixing vessel in turn and click on the button marked ">". You now have viscosity selected as the response and temperature and "mixing vessel" selected as the factors to include in the plots. The input screen should look like this:-

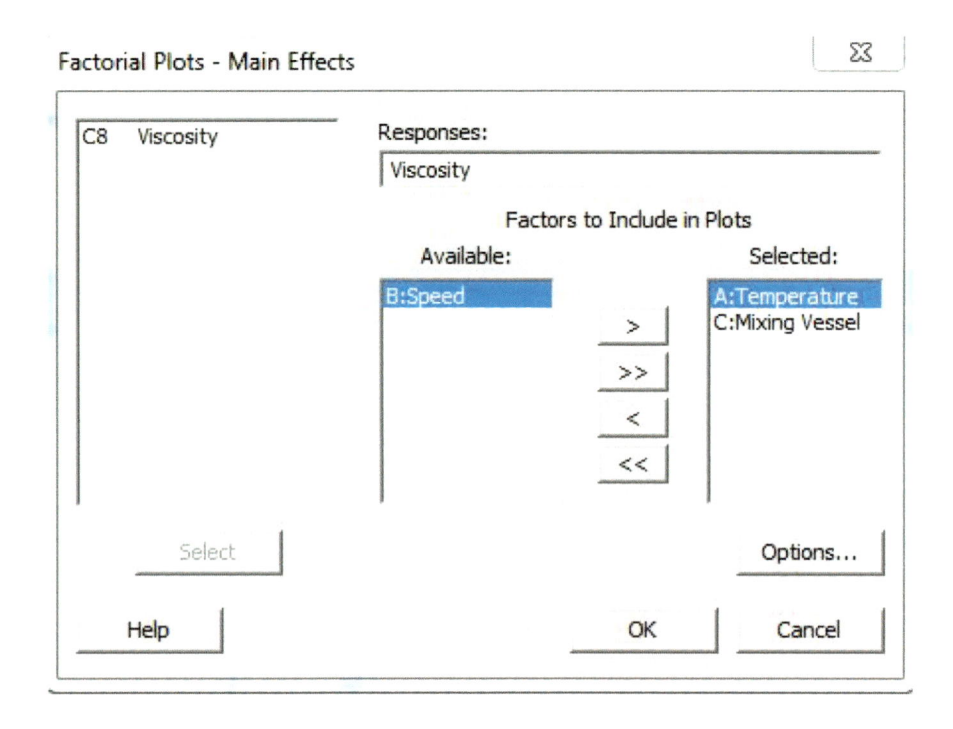

[Screenshot of Minitab "Factorial Plots - Main Effects" dialog box, showing C8 Viscosity in the variable list, Responses box containing "Viscosity", Available factors list showing "B:Speed" (highlighted), and Selected factors list showing "A:Temperature" and "C:Mixing Vessel". Buttons visible: >, >>, <, <<, Select, Options..., Help, OK, Cancel.]

You may find occasionally that the response variable, in this case, "viscosity", is greyed out and cannot be selected. If this is the case, click in the box marked "Responses" and the "Select" button will become live and enable you to select viscosity as the response.

The "Options" button simply allows you to define a title for the plots. We will ignore this for now.

Click on the button marked "OK" and then "OK" again on the remaining input screen. Minitab will now produce a main effects plot for viscosity. It should look like this: -

Main Effects Plot for Viscosity
Data Means

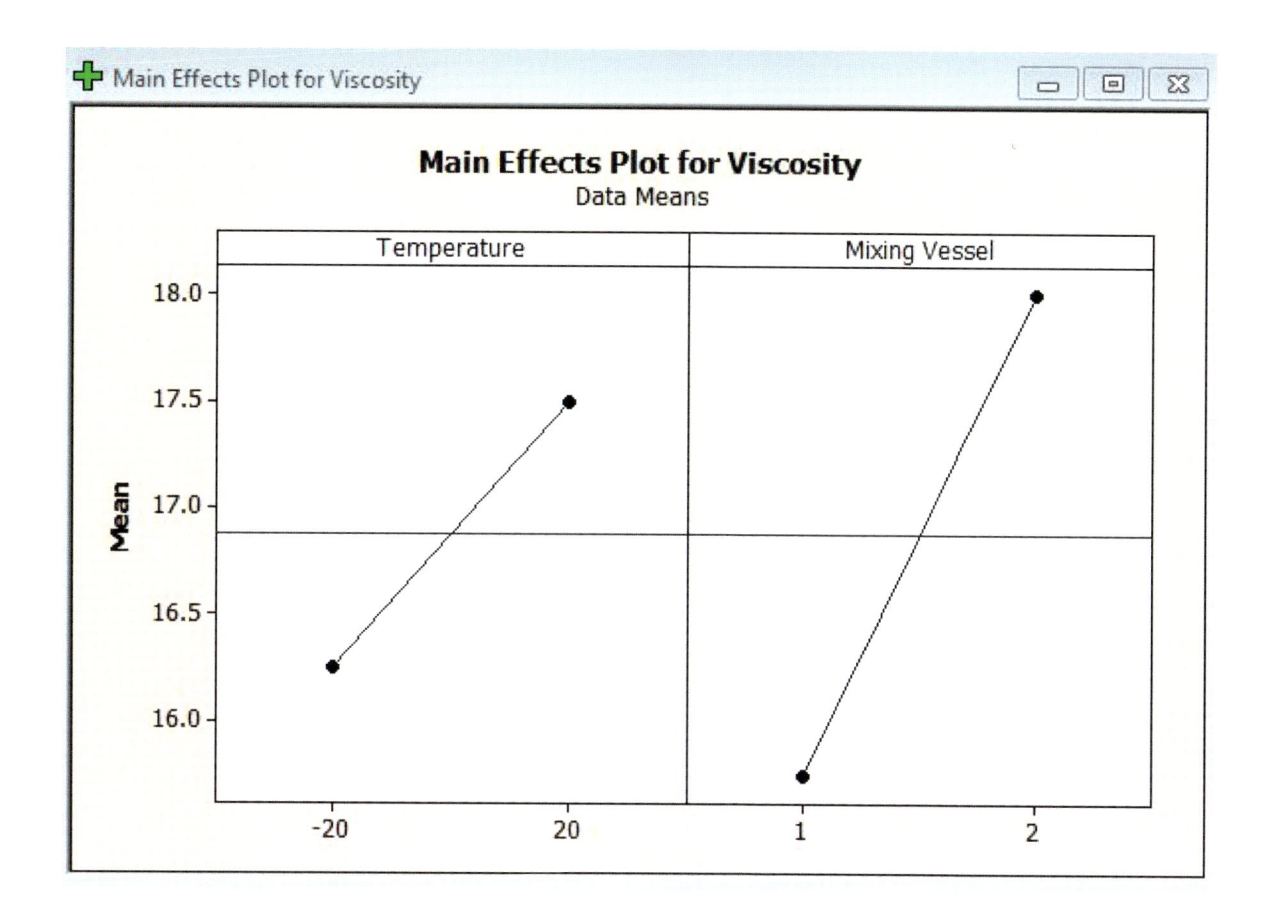

[Plot showing Temperature (-20 to 20) and Mixing Vessel (1 to 2) main effects on mean Viscosity. Both factors show increasing trends; Temperature rises from ~16.25 at -20 to ~17.5 at 20; Mixing Vessel rises from ~15.8 at 1 to ~18.0 at 2.]

You will remember that we are looking to make viscosity as low as possible at the lowest possible cost. You can see, from the plots above that mixing vessel number 1 will give the lowest viscosity and viscosity appears to be lower at a temperature of -20°C.

Now one thing to notice here is that the graph on the left should NOT be interpreted as meaning that viscosity and temperature have a linear relationship. Since we are only testing at two temperatures, Minitab can only assess the effect at those two temperatures and assume a straight line relationship between the two.

If we want to assess the relationship to see if any curvature exists in the relationship line then we will need to test at more than two levels.

So, we know that mixing vessel number 1 is the preferred vessel. Temperature is more difficult since the lower the mixing temperature the more it costs. In fact when we analyse the costs we discover that mixing at -20°C is three times the cost of mixing at +20°C. So what should we do?

The best approach is to carry out another experiment, this time using just mixing vessel number and at three temperatures.

Since we are starting to understand the relationship at the two extremes of temperature just tested, we can simply add in a test at another temperature. We can therefore try the mid-point of zero degrees.

An experienced operator has also come forward and proposed that length of mixing time may also have an effect. She has observed that when she mixes for a longer time then the viscosity appears to be lower. We decide to test this suggestion, along with temperature.

So we have the following proposed experiment: -

- We have two factors: temperature and mixing time.
- We have 3 levels for temperature, -20°C, 0°C and +20°C.
- We have 2 levels for mixing time.

The current mixing time is defined as between 2 minutes and 2 minutes 30 seconds. Since we don't know how viscosity changes with mixing time (apart from the anecdotal evidence from the experienced operator) we decide to test either side of the current mixing time.

How should we choose each level?

The general belief is that a very low mixing time will result in a poor mix and therefore there will be a minimum time required. We therefore decide to set the lower level of mixing time to 1 minute 30 seconds.

The safety case for the process defines that mixing time cannot exceed 5 minutes since a potentially dangerous reaction may occur in the mix constituents. We therefore decide to set the higher level of mixing time to 4 minutes.

So to recap; we will do a full factorial experiment with two factors. One factor has 3 levels and the other factor has two levels. Are you starting to become familiar with the terminology?

We decide we can do the experiment easily on one day and therefore blocking will not be required.

This number of factors and levels will require just 6 runs and the runs do not take much time to complete so we therefore decide to do 3 replicates. This is a total of 18 runs.

Use the previous screenshots to see if you can create the experiment.

You should get this: -

↓	C1 StdOrder	C2 RunOrder	C3 PtType	C4 Blocks	C5 Mixing Time	C6 Temperature	C7 Viscosity
1	7	1	1	1	90	-20	
2	2	2	1	1	90	0	
3	16	3	1	1	240	-20	
4	10	4	1	1	240	-20	
5	11	5	1	1	240	0	
6	4	6	1	1	240	-20	
7	14	7	1	1	90	0	
8	12	8	1	1	240	20	
9	5	9	1	1	240	0	
10	1	10	1	1	90	-20	
11	8	11	1	1	90	0	
12	13	12	1	1	90	-20	
13	3	13	1	1	90	20	
14	15	14	1	1	90	20	
15	6	15	1	1	240	20	
16	9	16	1	1	90	20	
17	18	17	1	1	240	20	
18	17	18	1	1	240	0	

As expected the operators ask if they can carry out all the runs at -20°C, then all the runs at 0°C and finally all the runs at 20°C. This will make it easier for them.

What is your response?

Let's look at the facts.

Removing randomisation

The experiment can be done in one day and on the same shift. The operators normally switch jobs half way through the shift to add variety to the day and therefore they propose that this is what they will do for the experiment.

The problem with this is that the operators also weigh and add the constituents of the mix, which may or may not have an effect. You therefore decide to agree to the request to do the three temperatures in order, but insist that the experiment (and therefore all runs) MUST be carried out by the same operator. This will eliminate any possible effect from changeover of operators. The operators agree once you have explained the reasons for the decision.

The experiment goes ahead, the results are collated and entered into our experimental design in Minitab. You can see the results entered in the following screenshot:

↓	C1 StdOrder	C2 RunOrder	C3 PtType	C4 Blocks	C5 Mixing Time	C6 Temperature	C7 Viscosity
1	7	1	1	1	90	-20	15
2	2	2	1	1	90	0	17
3	16	3	1	1	240	-20	14
4	10	4	1	1	240	-20	13
5	11	5	1	1	240	0	13
6	4	6	1	1	240	-20	15
7	14	7	1	1	90	0	16
8	12	8	1	1	240	20	17
9	5	9	1	1	240	0	13
10	1	10	1	1	90	-20	16
11	8	11	1	1	90	0	18
12	13	12	1	1	90	-20	16
13	3	13	1	1	90	20	17
14	15	14	1	1	90	20	18
15	6	15	1	1	240	20	16
16	9	16	1	1	90	20	17
17	18	17	1	1	240	20	15
18	17	18	1	1	240	0	13

If you carry on through the analysis you will notice something strange. The option to select normal, half normal and Pareto effects plots is not available.

This is because they are only made available for two level factorial experiments and not general full factorial experiments with 3 or more levels. The explanation for this will make your brain hurt, but I have provided such an explanation in Appendix 1. I have also reproduced an explanation of the plotted residual options offered by Minitab.

To keep it simple, here is what I recommend:

Click on **Stat/DoE/Factorial/Analyse Factorial Design/Graphs**

You'll see this:-

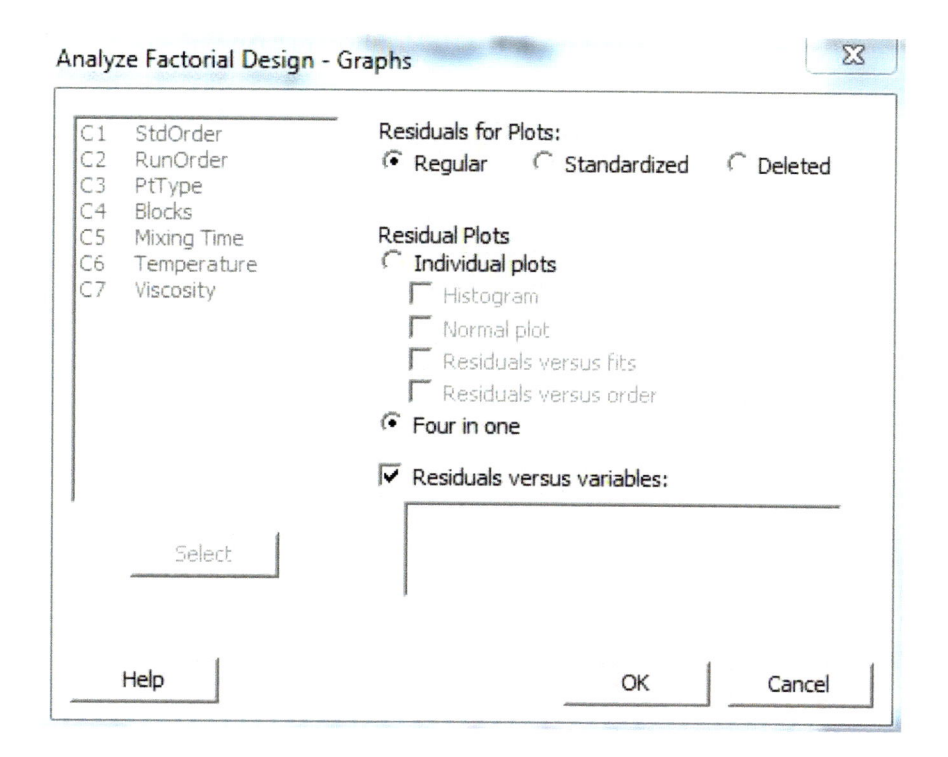

Now select "Standardized" and the "Four in One" options. Click "OK"

Minitab will then produce the required plots as shown below: -

Residual Plots for Viscosity

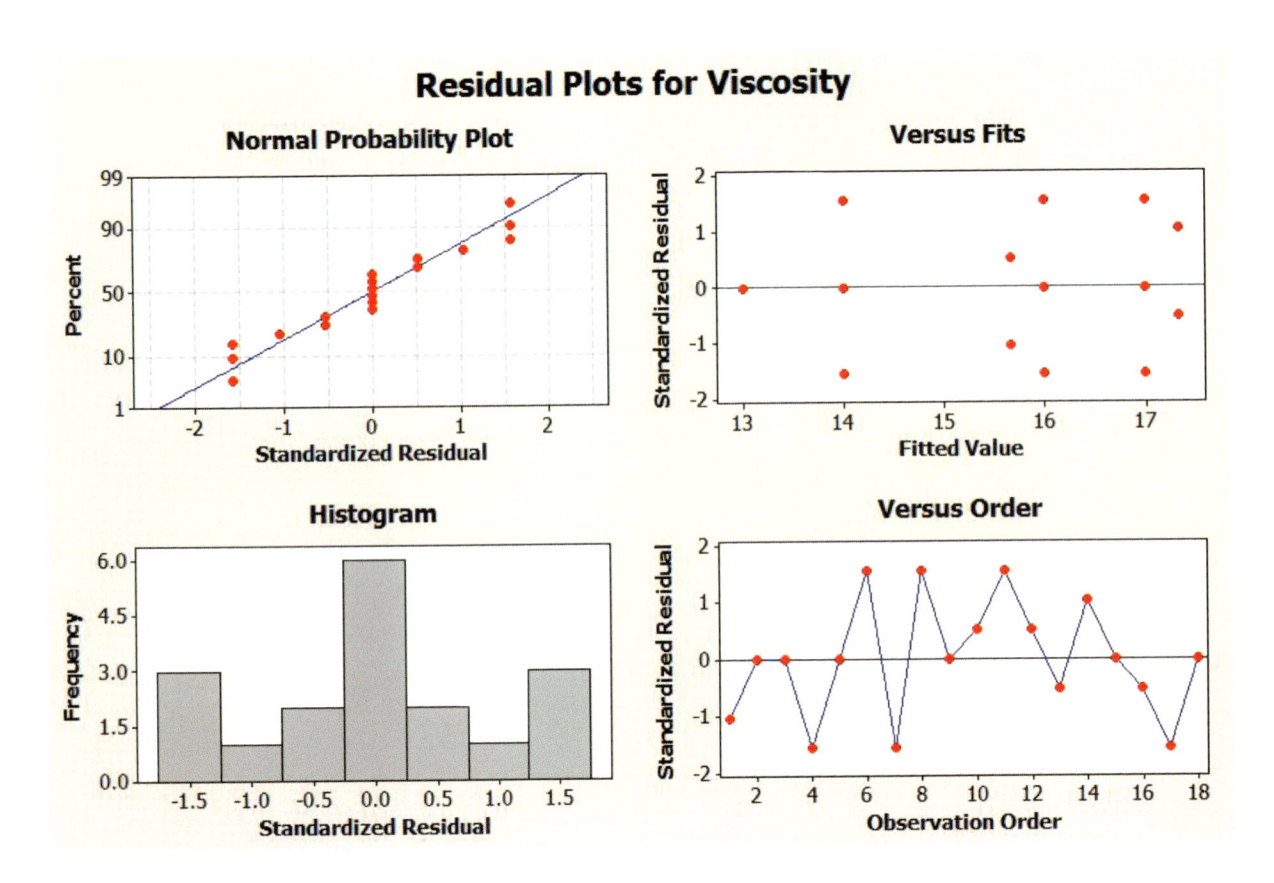

Let's examine what these tell us.

The Normal Probability plot and the Histogram plot: The data looks generally straight, with no bars far from the others, so we can assume normality with no outliers.

The Versus fit plot: the data looks random around zero with no increase in spread as the fitted values increase. This indicates that the "constant variance" assumption is valid. Don't worry – it's a good thing!

The Versus order plot: this is a plot of data in collection order, and we are looking for non-random error particularly with time. Looking at the results it appears random with no real patterns.

The next step is to plot the main effects since you'll remember that in the first experiment we looked at the effects of temperature and mixing vessel. Unfortunately those results indicated that a lower temperature resulted in the required reduction in viscosity, and at a temperature of -20°C we had discovered that the cost of mixing was 3 times that of mixing at 20°C.

An experienced operator then suggested that mixing time seemed to have an effect on viscosity so we decided to test this suggestion, together with the temperature at 3 levels to see if the relationship was linear.

Plotting the main effects shows an interesting response:

Main Effects Plot for Viscosity
Data Means

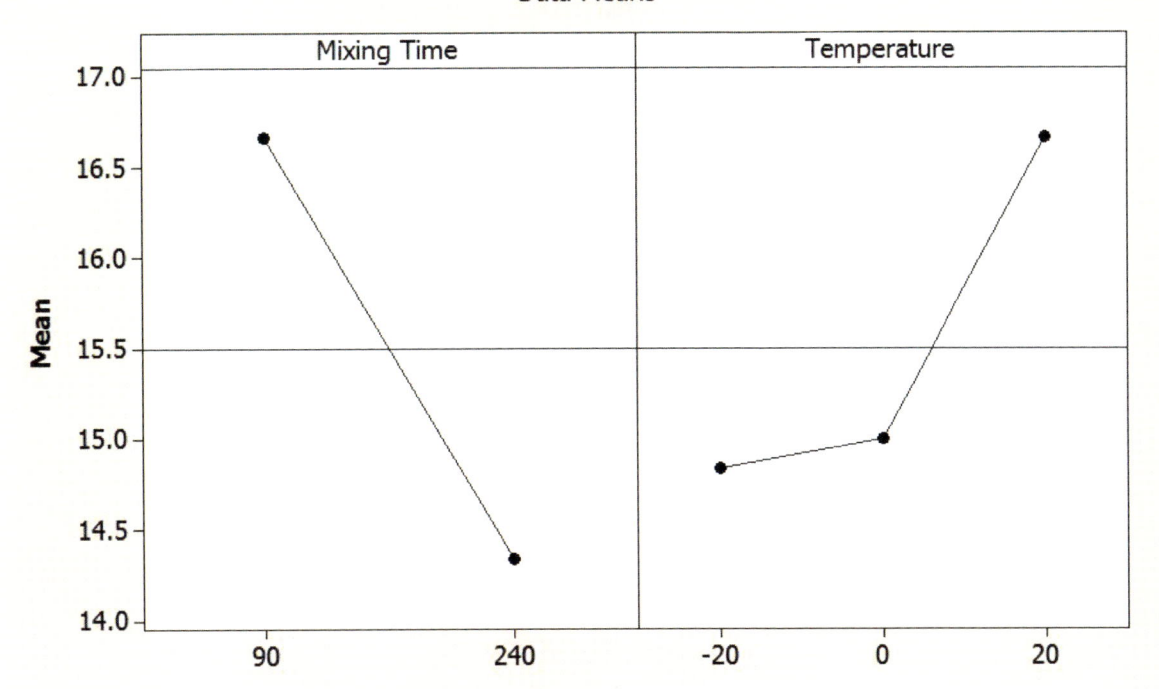

You can see that the operator was right and an increased mixing time does indeed reduce the viscosity. The effect of temperature looks to be minimal between -20°C and 0°C, but becomes much more significant between 0°C and 20°C. The graph suggests that we can mix at zero degrees (and maybe higher) for 240 seconds (i.e. 4 minutes) and get a viscosity around 14.5.

It is possible that we could run another limited experiment to assess the temperature effect between 0°C and 20°C since we don't know the temperature at which the viscosity starts to rise. We only know that it is between 0°C and 20°C. The decision on whether to do this or not will depend on the costs of mixing. If the cost of mixing at 0°C is low enough in comparison with the cost at -20°C then it is acceptable to mix at 0°C.

Going back and doing an interaction plot, i.e. a plot of mixing time and temperature together, generates the following : -

Interaction Plot for Viscosity
Data Means

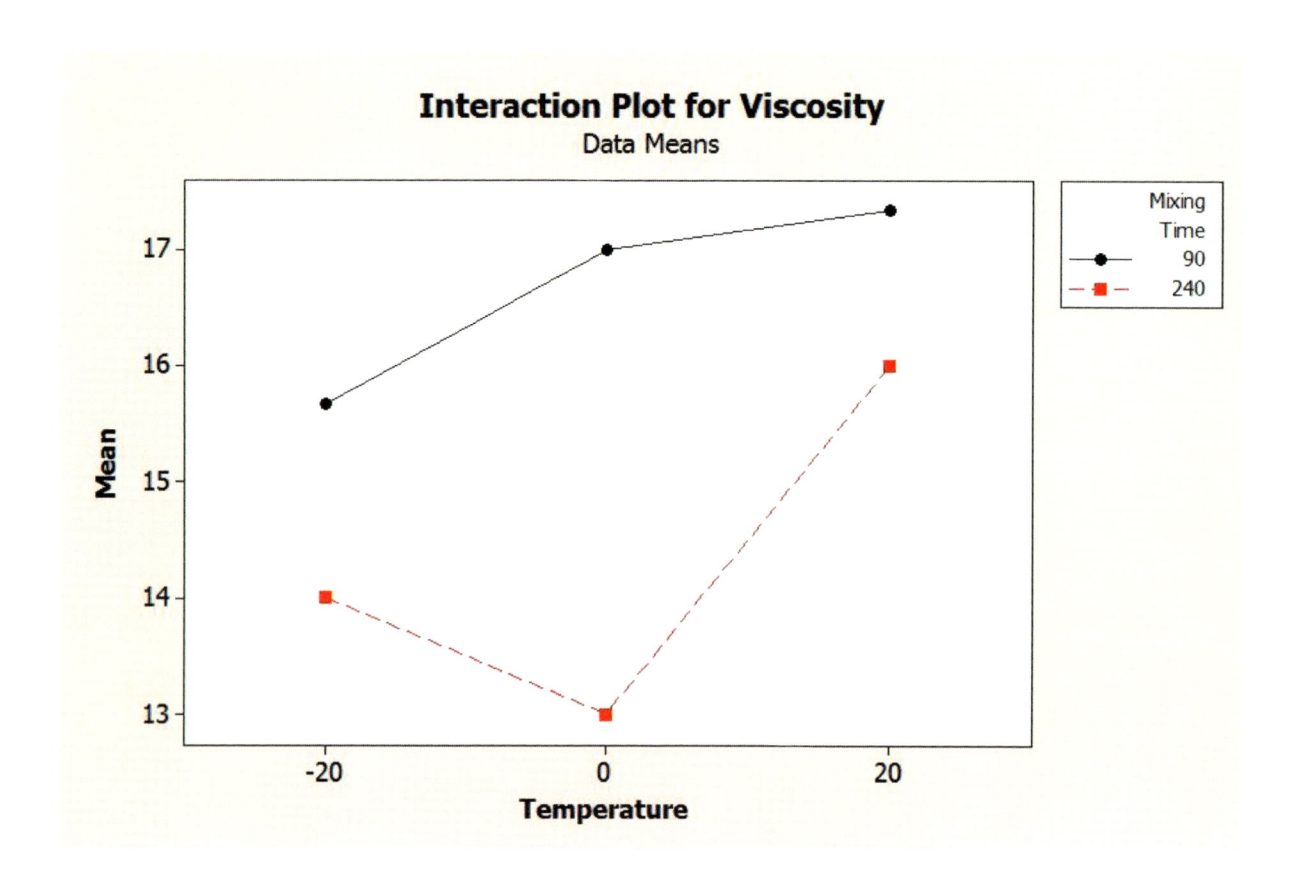

This shows the effect of temperature at an extended mixing time of 240 seconds is far greater than that for a 90 second mixing time.

The optimal set-up is to mix for 240 seconds at a temperature of 0°C. This gives the lowest viscosity levels.

Ok, that's a very practical demonstration of the power of a properly designed experiment!

A Worked Example

Let's now turn our attention to a worked example.

In this example I will limit the number of screenshots since you have seen examples of them in the earlier text. I will concentrate instead on the thought processes which I go through, particularly when analysing the results of an experiment.

Let's look at a typical example of a problem where DoE is an appropriate and powerful tool in finding the solution.

So, let's imagine that a vehicle manufacturer needs to understand how best to prevent rippling of paint applied to his vehicles.

DoE can be used to test the effect of these variables and, where possible, reduce the temperature requirement since heating of either paint or the vehicle surface is expensive.

Identification of factors, levels and responses

The paint experts within the company have identified three factors which may affect rippling: these are temperature of the surface to which the paint is applied, temperature of the paint, and the speed of application. The actual sweeping speed of the paint gun will be kept constant.

The current settings are:

- Temperature of the surface: - 20°C
- Temperature of the paint: - 27°C
- Volume of paint at the nozzle: - 120cc per minute

Since the first step is to get an indication of the size of each of these effects we decide on the following levels of each factor:

- Temperature of the surface; lower level set to 5°C and upper level set to 30 °C
- Temperature of the paint; lower level set to 15°C and upper level set to 40 °C
- Volume of paint at the nozzle; lower level set to 60cc per minute and upper level set to 180cc per minute.

Note that the settings are as wide as we can make them without compromising safety. This allows effective quantification of the size of each effect.

The response is simply the amount of paint rippling. How best should we measure it? Since this is a difficult response to measure objectively we will use the most experienced QA inspector to rate the amount of rippling from each run of the experiment on a scale of 1 to 5, with 0 representing no visible ripples and 5 representing a significant amount of rippling.

The same inspector will judge all runs of the experiment to reduce bias and he will not be told which run is which to prevent any effect of prior knowledge of the process.

We have decided not to use blocking since we can do all runs quite quickly, on the same shift and with the same operator. We have also decided we will have two replicates.

Creating the Design

We can quickly see that we have three factors, two levels of each and one response. We know the level settings so we can now create the design. If needed go back to the previous sections on creating the design and refresh yourself on how to do this. Then have a go – you should end up with something like this: -

↓	C1 StdOrder	C2 RunOrder	C3 CenterPt	C4 Blocks	C5 Surface Temperature	C6 Paint Temperature	C7 Volume of Paint
1	16	1	1	1	30	40	180
2	1	2	1	1	5	15	60
3	2	3	1	1	30	15	60
4	8	4	1	1	30	40	180
5	9	5	1	1	5	15	60
6	11	6	1	1	5	40	60
7	14	7	1	1	30	15	180
8	4	8	1	1	30	40	60
9	7	9	1	1	5	40	180
10	12	10	1	1	30	40	60
11	6	11	1	1	30	15	180
12	3	12	1	1	5	40	60
13	5	13	1	1	5	15	180
14	10	14	1	1	30	15	60
15	15	15	1	1	5	40	180
16	13	16	1	1	5	15	180

Note that you may have a different run order since Minitab randomises this but the standard order should be the same parameters for each. Be careful when entering the results that the correct result is put in against the correct run parameters for that run. In the case above one of the runs is for standard order 7 with a surface temperature of 5, a paint temperature of 40 and volume of paint at 180. If you look at your design then the standard order for run 7 (as an example) should have the same settings, no matter where it appears in the run order.

Running the Experiment

The experiment is carried out using the runs listed in the design and the results collated. Care should be taken to ensure that all the settings are as described and ideally the experiment should be carried out in the run order shown. The results are entered into the design and are as shown in the table below.

↓	C1	C2	C3	C4	C5	C6	C7	C8
	StdOrder	RunOrder	CenterPt	Blocks	Surface Temperature	Paint Temperature	Volume of Paint	Paint Ripple Score
1	16	1	1	1	30	40	180	5
2	1	2	1	1	5	15	60	2
3	2	3	1	1	30	15	60	2
4	8	4	1	1	30	40	180	4
5	9	5	1	1	5	15	60	1
6	11	6	1	1	5	40	60	1
7	14	7	1	1	30	15	180	5
8	4	8	1	1	30	40	60	2
9	7	9	1	1	5	40	180	4
10	12	10	1	1	30	40	60	3
11	6	11	1	1	30	15	180	5
12	3	12	1	1	5	40	60	3
13	5	13	1	1	5	15	180	4
14	10	14	1	1	30	15	60	2
15	15	15	1	1	5	40	180	3
16	13	16	1	1	5	15	180	4

Analysing the Results

Ok, so we have the results and we can now proceed with the analysis. The first step is to see which of the responses are significant. Click on **Stat/DOE/Factorial/Analyse Factorial Design,** as described in the previous sections, and select "Paint Ripple Score" as the response.

Click on "Graphs", select the Normal Plot and the Pareto Plot, together with "Standardized" and "Four in one". Click "Ok" twice on the respective screens and then Minitab will produce the three plots you selected. In this case, the three plots are: the Normal plot of the standardized effects, the Pareto plot, and the four in one plot. I have reproduced the plots below: -

Pareto Chart of the Standardized Effects
(response is Paint Ripple Score, Alpha = 0.05)

Factor	Name
A	Surface Temperature
B	Paint Temperature
C	Volume of Paint

Normal Plot of the Standardized Effects
(response is Paint Ripple Score, Alpha = 0.05)

Effect Type:
- Not Significant
- Significant

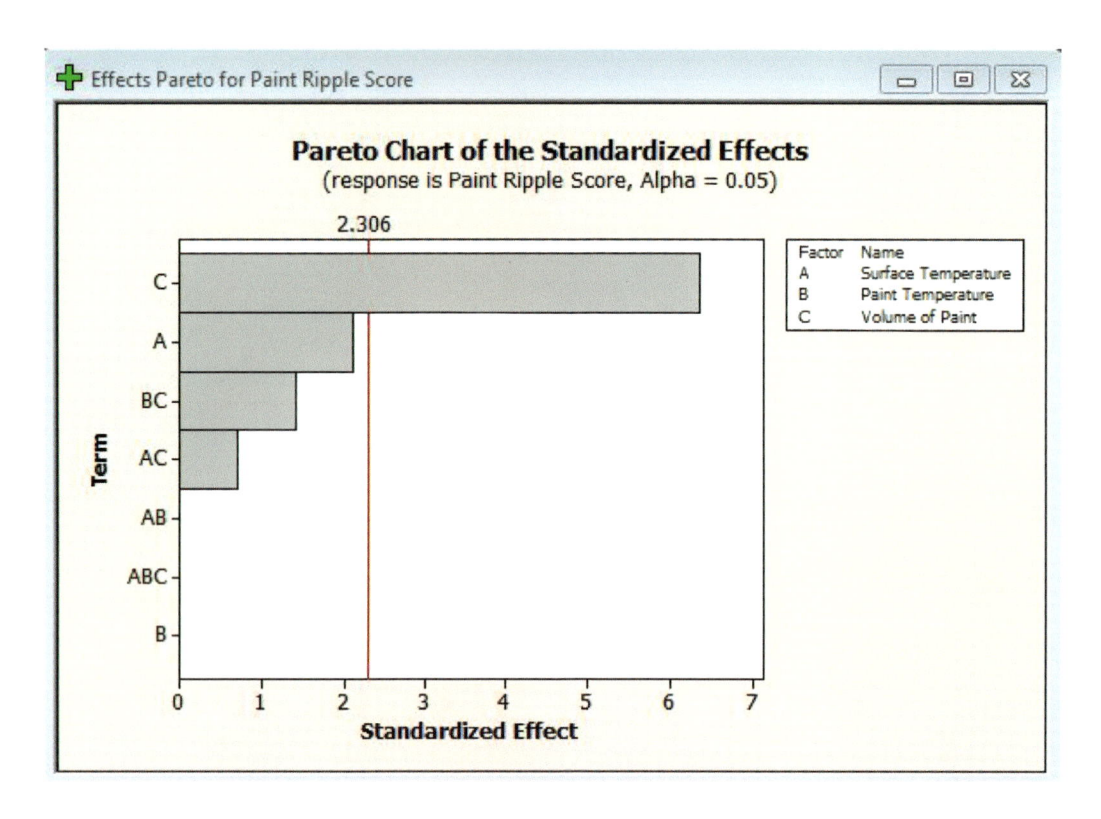

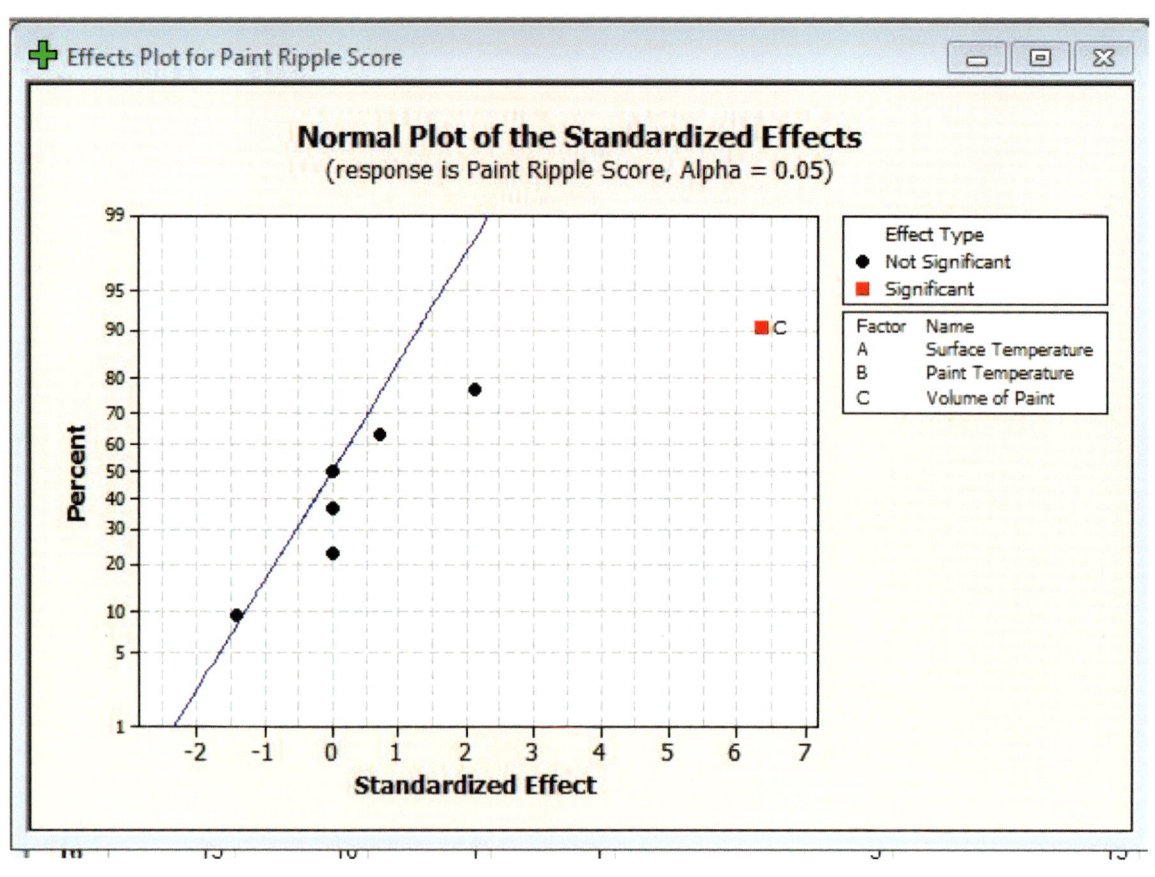

Factor	Name
A	Surface Temperature
B	Paint Temperature
C	Volume of Paint

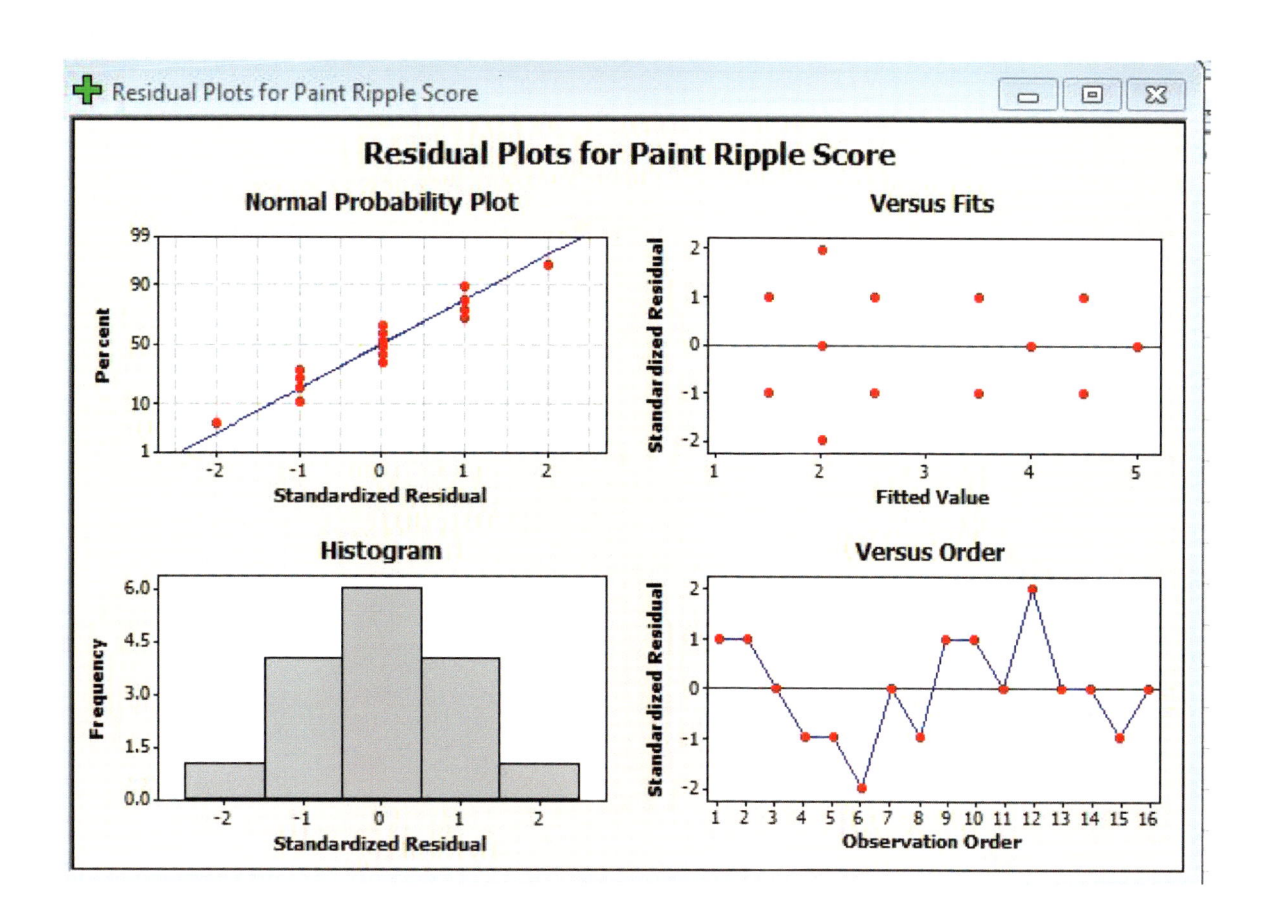

Residual Plots for Paint Ripple Score

If you look at the Pareto plot then you can see that factor C, volume of paint, is highly significant. With factor A, surface temperature, close to being significant at an alpha level of 0.05.

In this case, we can go back and change to an alpha level of 0.10 instead of 0.05 and see if the effect becomes significant. We repeat the previous step of producing the three graphs and the resulting Pareto and Normal Plot of the standardised effects does show that factor A has now become significant.

We now can look at the main effects plots. To do this, click on **Stat/DOE/Factorial/Factorial Plots.**

Now select "main Effects Plots and Click on "Setup" to select the response, "Paint Ripple Score" and the factors to include in the plots. We only really need to select the two effects that are significant (i.e. surface temperature and volume of paint), but for completeness we'll include all three. The resulting graph looks like this: -

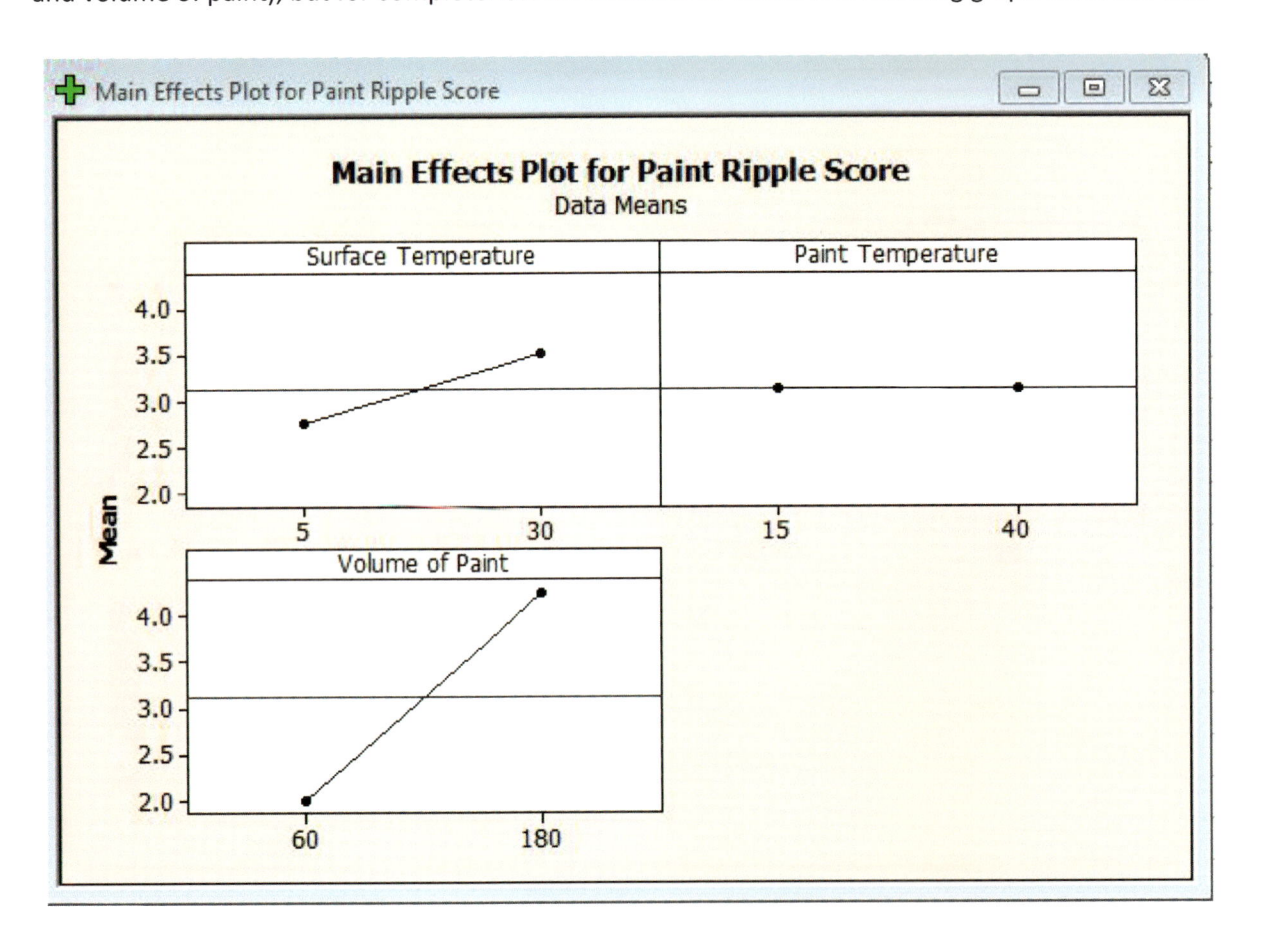

Our conclusion that paint temperature is not significant is clearly correct since the plot shows it as a straight horizontal line.

With these types of plots, the greater the slope of the line, the greater the effect.

We can see that an increased volume of paint results in a significantly worse score for rippling (remember that a score of zero was "no visible rippling" and a score of 5 was "a significant amount of rippling").

We can also see that a lower surface temperature reduces the observed rippling effect.

We now need to consider if we can reduce the rippling effect further, perhaps by reducing the volume of paint (which will also save money on paint quantity used) and reducing the surface temperature. Clearly this will depend on the cost of reducing the surface temperature. After some discussion we discover that reducing the surface temperature from 5 degrees to zero degrees is only marginally more expensive and the cost of further paint quantity reduction will more than offset this increased cost.

It would seem sensible to carry out a further experiment with two factors at two levels, and three replicates to assess the variability of the response.

The factors and levels would be: -

- Temperature of the surface: - lower level set to 0°C and upper level set to 5 °C
- Volume of paint at the nozzle: - lower level set to 40cc per minute and upper level set to 60cc per minute.

The experiment is created and looks like this: -

	C1	C2	C3	C4	C5	C6
	StdOrder	RunOrder	CenterPt	Blocks	Temperature of the surface	Volume of paint
1	8	1	1	1	5	60
2	1	2	1	1	0	40
3	7	3	1	1	0	60
4	3	4	1	1	0	60
5	12	5	1	1	5	60
6	2	6	1	1	5	40
7	5	7	1	1	0	40
8	11	8	1	1	0	60
9	4	9	1	1	5	60
10	10	10	1	1	5	40
11	9	11	1	1	0	40
12	6	12	1	1	5	40

We run the experiment and the results are entered, as shown below: -

	C1	C2	C3	C4	C5	C6	C7
	StdOrder	RunOrder	CenterPt	Blocks	Temperature of the surface	Volume of paint	Paint Ripple Score
1	8	1	1	1	5	60	2
2	1	2	1	1	0	40	1
3	7	3	1	1	0	60	2
4	3	4	1	1	0	60	3
5	12	5	1	1	5	60	2
6	2	6	1	1	5	40	2
7	5	7	1	1	0	40	1
8	11	8	1	1	0	60	2
9	4	9	1	1	5	60	2
10	10	10	1	1	5	40	2
11	9	11	1	1	0	40	1
12	6	12	1	1	5	40	2

The normal plot and Pareto plot are also produced:

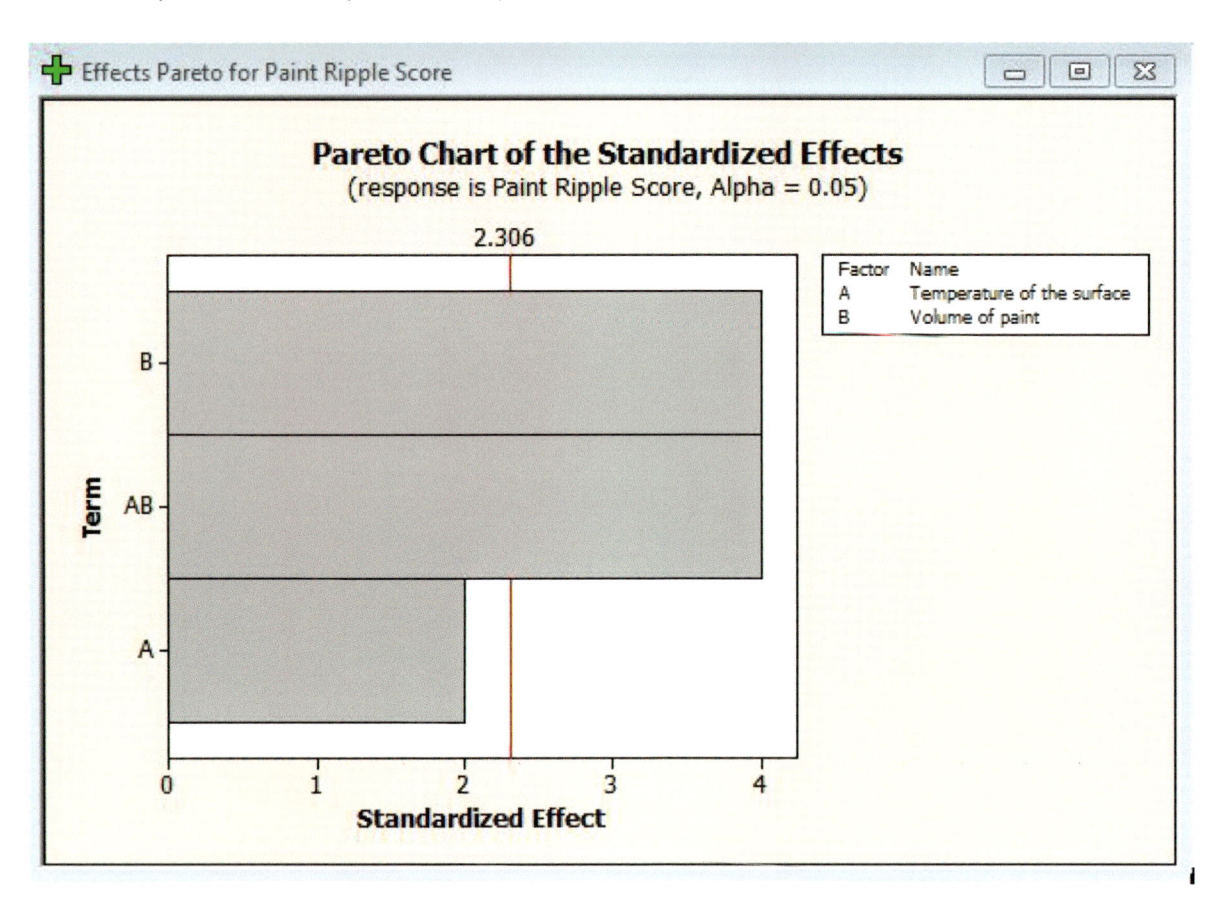

Effects Pareto for Paint Ripple Score

Pareto Chart of the Standardized Effects
(response is Paint Ripple Score, Alpha = 0.05)

2.306

Factor	Name
A	Temperature of the surface
B	Volume of paint

Terms (top to bottom): B, AB, A

Standardized Effect (x-axis): 0, 1, 2, 3, 4

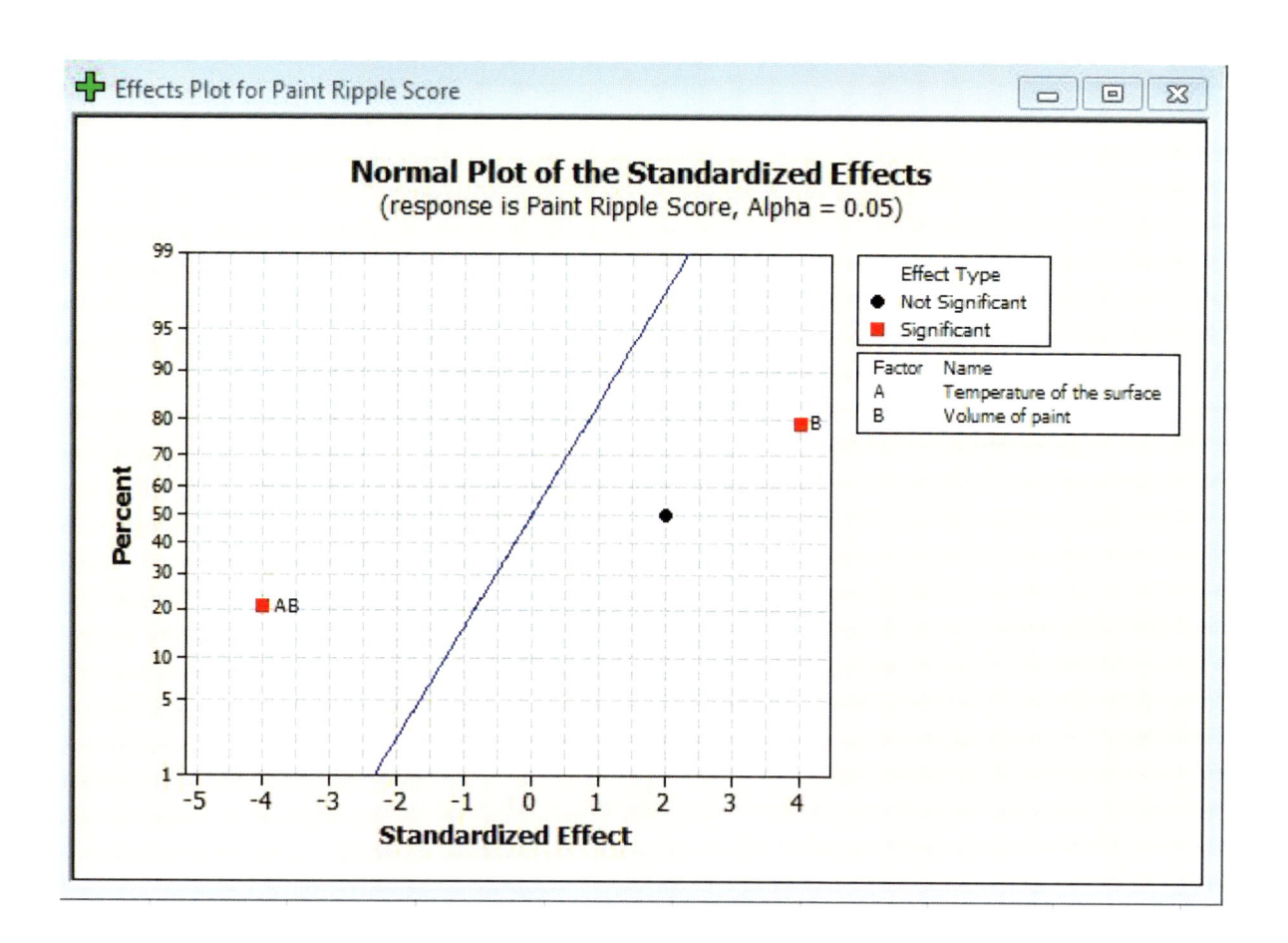

Normal Plot of the Standardized Effects
(response is Paint Ripple Score, Alpha = 0.05)

We can see from the plots that two factors are significant at an alpha level of 0.05. These are the main factor B, the volume of paint and the interaction AB, the volume of paint combined with the temperature of the surface. Since we have an interaction term which is significant, let's examine the interaction plot:

Interaction Plot for Paint Ripple Score

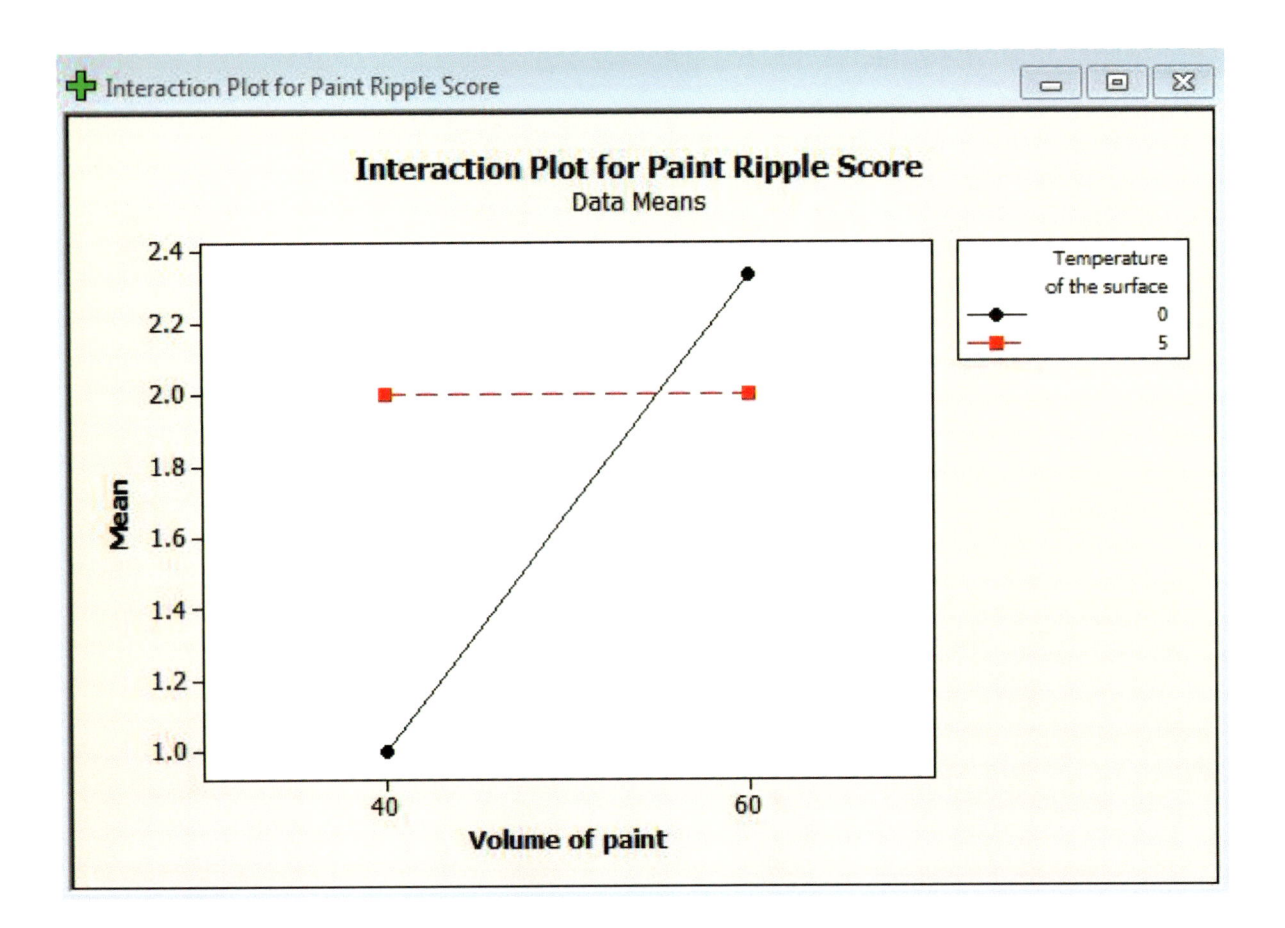

We can see from this plot that **at a surface temperature of 5°C** the volume of paint has no effect. It is also apparent that reducing the surface temperature to zero degrees results in the volume of paint becoming highly significant. The best results (i.e. a mean ripple score of 1.0) are achieved when we have volume of paint set to 40cc and the surface temperature at 0°C.

We have already ascertained that the cost of reducing the surface temperature to 0°C is little more than a surface temperature of 5°C and that the savings from a reduced volume of paint will offset this small extra cost. We therefore have a solution to maximise the quality of paint finish and minimise the rippling!

Summary

Ok, that's it. In this short guide to Design of Experiments, I've tried to keep things as simple and practical as possible.

Some of the methods I have described may upset the purist statisticians and mathematicians out there, but I've used these methods over more than thirty years to find solutions to seemingly insolvable problems resulting in savings of tens of millions of pounds. To me that's proof enough that I'm doing something right!

DoE is a very powerful technique to solve problems far quicker than changing one factor at a time and then testing the result, where such a technique will never identify interactions between factors and such interactions are very common in engineering and science. Try the techniques described in this guide and you'll see what I mean.

Glossary of Terms

Factor: Parameters which are deliberately varied in an experiment in order to determine their effect on one or more responses.

Response: Parameters which vary as a result of factor(s)

Level: The chosen settings for each factor. These are normally set as wide as safely possible so that the effect of each factor can be assessed.

Run: A set of experimental conditions in which each of the factors is held at a specific level.

Full Factorial: An experiment in which every combination of factors is tested. Fractional factorial experiments do not test every available combination.

Confounding: This means that some of the main effects and interactions cannot be estimated independently from each other. Confounding therefore cannot exist in full factorial experiments since every combination is tested resulting in all main effects and interactions being able to be estimated independently.

Randomizing: This is used to balance the effect of uncontrollable factors which can affect the responses of an experiment. Runs should usually be performed in a random order since this will reduce the probability that such uncontrollable factors bias the results of an experiment.

Interactions: Occurs where two or more factors acting together on a response. Two way interactions exist where two factors act together, three way interactions exist where three factors act together, etc.

Alias: An alias structure describes the confounding which exists in a specific designed experiment.

Resolution: Describes the extent of aliasing of effects in a fractional factorial design. If designing a fractional factorial experiment, it is generally best to choose design with the highest available resolution.

Replicate: Is a duplicate set of complete runs from the complete design.

Block: A set of runs in the designed experiment which are all carried out at as near constant conditions as possible, other than those being varied as part of the experiment. Using blocks is typically used to minimise bias due to these extraneous factors. The run order should be randomize within any blocks.

Covariate: It used to account for a factor that is difficult to control and to reduce error variance.

Alpha Level: Is a measure of the risk of rejecting a true null hypothesis. A null hypothesis is one in which you assert that the factors being tested are unrelated and that therefore the results are the result of random events of chance. The level is expressed as a probability between 0 and 1.

Residual: Is the difference that exists between a fitted value in a particular model and an observed value.

Appendix 1: Normal Plots of the Effects for Multilevel Factorial Designs

Two level designs using standardized effects are provided to select significant terms in the design. General factorial designs use normalized effects instead. Minitab offers four different residual plots for regular, standardized or deleted residuals.

You can see these in the screenshot below:

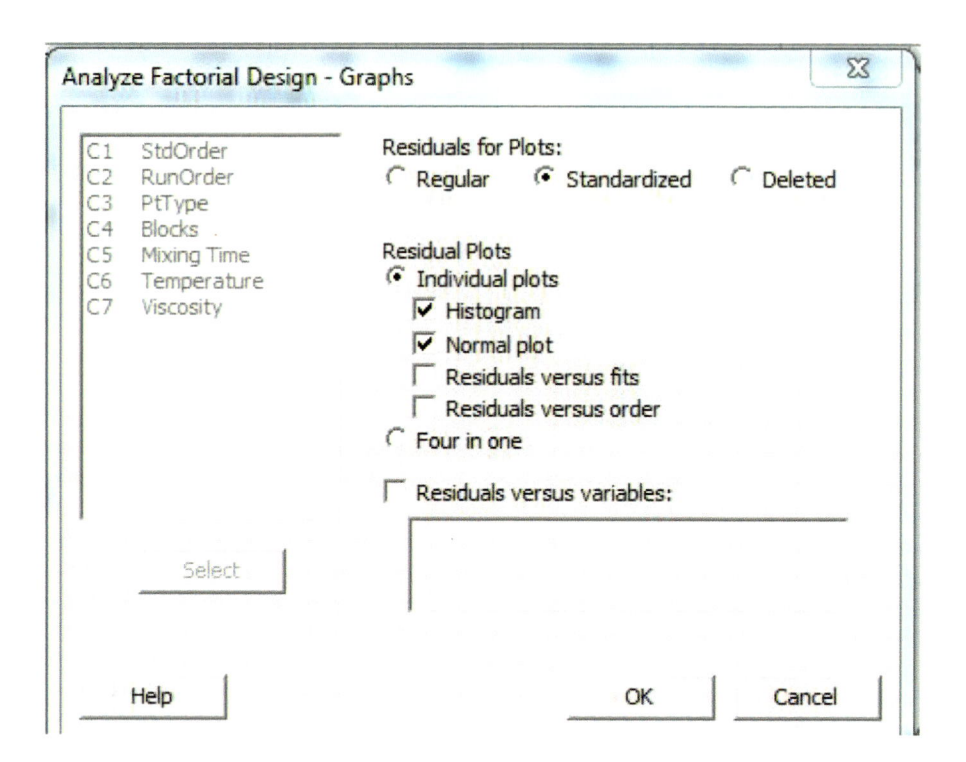

In this screenshot you can see that only two residual plots are selected, but you can choose all four (the four in one option) in addition to the "Residuals versus variables" option.

A residual is simply the difference between an observed value (let's call it "y") and its corresponding fitted value (let's call it "z"). If you draw a scatter plot of one variable against another then a regression line can be plotted showing the fitted values of one variable against the observed value for the other variable. (I told you it would make your brain hurt!).

In real life the fitted values will usually lie both above and below the regression line. Residuals are useful because they tell you to what extent a model accounts for the variation in the observed data.

Regular residual plots should be used when you wish to examine the residuals in the original scale of the data.

Standardized and deleted residuals are useful for detecting outliers in a data set. They are superior to regular residuals in indicating outliers.

The plotted residuals options offered by Minitab are described below and are direct quotes from Minitab's help facility, courtesy of Minitab Inc. Such material remains the exclusive property and copyright of Minitab Inc.

Histogram of the Residuals - This is an exploratory tool to show general characteristics of the residuals including typical values, spread, and shape. A long tail on one side may indicate a skewed distribution. If one or two bars are far from the others, those points may be outliers.

Normal Probability Plot of residuals - The points in this plot should generally form a straight line if the residuals are normally distributed. If the points on the plot depart from a straight line, the normality assumption may be invalid.

Residuals Versus Fitted Values - This plot should show a random pattern of residuals on both sides of 0. If a point lies far from the majority of points, it may be an outlier. There should not be any recognizable patterns in the residual plot. For instance, if the spread of residual values tend to increase as the fitted values increase, then this may violate the constant variance assumption.

Residuals Versus Order of Data - This is a plot of all residuals in the order that the data was collected and can be used to find non-random error, especially of time-related effects. This plot helps you to check the assumption that the residuals are uncorrelated with each other.

Residuals versus predictors - This is a plot of the residuals versus a predictor. This plot should show a random pattern of residuals on both sides of 0. Non-random patterns may violate the assumption that predictor variables are unrelated to the residuals. You may have used an incorrect functional form to model the curvature.